THE CITY

DYNAMICS OF URBAN PROPERTY DEVELOPMENT

URBAN ECONOMICS

DYNAMICS OF URBAN PROPERTY DEVELOPMENT

JACK ROSE

LONDON AND NEW YORK

First published in 1985

This edition published in 2007
Routledge

2 Park Square, Milton Park, Abingdon, Oxfordshire, OX14 4RN
270 Madison Avenue, New York NY 10016

Routledge is an imprint of Taylor & Francis Group, an informa business

First issued in paperback 2010

British Library Cataloguing in Publication Data
A CIP catalogue record for this book
is available from the British Library

Dynamics or Urban Property Development

ISBN 13: 978-0-415-41318-3 (set)
ISBN 13: 978-0-415-41932-1 (subset)
ISBN 13: 978-0-415-41761-7 (volume) (hbk)
ISBN 13: 978-0-415-61147-3 (volume) (pbk)

Routledge Library Editions: The City

The Dynamics of Urban Property Development

Jack Rose

M.Phil., Faculty of Urban and Regional Studies,
University of Reading
Fellow of the Incorporated Society of Valuers and Auctioneers (ISVA)
Sometime Examiner in Valuations for the ISVA
Chairman, Land Investors PLC

LONDON NEW YORK
E. & F. N. SPON

First published 1985 by
E. & F. N. Spon Ltd
11 New Fetter Lane, London EC4P 4EE

Published in the USA by
E. & F. N. Spon
733 Third Avenue, New York NY 10017

Printed in Great Britain
at the University Press, Cambridge

ISBN 0 419 13160 4

British Library Cataloguing in Publication Data

Rose, Jack
The dynamics of urban property development.
1. Real estate—Great Britain
I. Title
333.3′8 HD598

ISBN 0–419–13160–4

Library of Congress Cataloging in Publication Data

Rose, Jack, 1961–
The dynamics of urban property development.

Bibliography: p.
Includes index.
1. Land use, Urban—Great Britain. 2. Real estate development—Great Britain. 3. City planning—Great Britain. 4. Urbanization—Great Britain. 5. Urban policy—Great Britain. 6. City planning and redevelopment law—Great Britain. I. Title.
HD596.R67 1985 333.77′15′0941 84–26717
ISBN 0–419–13160–4

To the memory of Philip Edward Rose

Contents

Preface

The history of urban development since the Industrial Revolution is best understood in the light of the general statement that economic growth, in changing the pattern of production and leading to the movement of a predominantly agricultural population from country to town, made it inevitable that socio-economic changes produced consequent changes in philosophical and political thinking. Not the least of these was the emergence, or rather the resuscitation of the opposing views of Plato and Aristotle expressed some 2000 years ago. In simple terms, Plato advocated that all property should be controlled by the State, whilst Aristotle dismissed his plans with the comment 'if they were good, someone would have thought of them sooner'.

Over two hundred years of virtually unabated development under conditions first of *laissez-faire*, then of ever increasing statutory controls, many expediently dictated by wars, financial booms and recessions, call for a study not only of the *history*, but the *issues* that should determine the future of the development and redevelopment of the built environment.

This book is about the dynamics of urban development and is based on a thesis written for a research degree at the Faculty of Urban and Regional Studies of the University of Reading.

The dictionary describes dynamics as 'that branch of mechanics treating the motion of bodies acted on by forces of any kind in any sphere, physical or moral.' The body is the quantum of urbanization that has occurred since the Industrial Revolution. The physical and moral forces are socio-economics and politics. The motion has persistently accelerated and is seemingly perpetual. What needs to be answered are the questions: in what direction and at what speed is to be the future progress of urban development?

Changes in the economy, the numbers and location of the population, taxation, planning and the landlord and tenant relationship have been the subject of extensive legislation and of numerous books, learned articles and media commentary. The purpose of this book is to encapsulate the more significant factors that have marked the progress of urban property development from the Industrial Revolution to the present time.

Whilst much of the material has been gleaned from records and published histories, that of the period covering the last fifty years is a personal impression and although a conscious effort has been made to be objective, no apology is offered if a subjective view emerges.

JACK ROSE 1985

Acknowledgements

First thanks are due to Professors Miles, Millington and George Herd of the University of Reading who in turn acted as supervisors of the original thesis on which this book is based. I am no less indebted to the librarians of London University, the British Museum, the Royal Institution of Chartered Surveyors and the Royal Town Planning Institute who made available so much of the historical information researched and to Ravenseft Properties Limited, J. Sainsbury PLC and W. H. Smith & Son Limited who provided the comprehensive details of their operations.

To acknowledge individually the authors of the many books referred to would be invidious, and to any one of those who recognizes a direct quote or the lifting of a reference he may have originally researched I offer the excuse that plagiarism is the art of faithfully copying one author's work; research is the term applicable to the copying of many.

Frances Rose typed the original thesis and Pat Norris the book in its present form, I owe both these ladies my grateful thanks. Not least of those who gave me valuable assistance is Dr Peter Ambrose of Sussex University whose advice on the format and content has been invaluable.

1

Introduction

1.1 THE ISSUES

Clearly, the most important issue is how best the development and redevelopment of the built environment should be managed, given changes both observed and predicted which include:

Population trends;
Technology of production;
Leisure and entertainment requirements resulting from increased leisure time;
Expectation of increased standards of living;

The need to replace outworn structures yet preserve historic and architecturally desirable buildings, both singly and in groups;
Attitudes to profits arising from development and the distribution between landowners, developers, financial institutions and the State in its role as tax collector.

At no time in the past has so much publicity been given to development and the debate centres on apparently conflicting needs. On the one hand there is the need for private capital to secure a competitive return on its investment. On the other there are the needs of the community (through the agency of the State) for schools, hospitals, and offices when some 60% of the country's entire work-force is employed by a Government with a limited capacity to raise and invest public monies. It is this movement of the frontier between the private and public sector, observed over the last fifty years, that provides the basis of the dialectic — is the present mix of private and public activity the best that can be devised given the need for certain objectives that all, or most, would agree to? Or does the balance lie in the direction of either greater or less State control?

1.2 THE HISTORY

Any historical study must involve a specific approach, in this case, socio-economic. H. T. Buckle in his *History of Civilization in England*[1], written in 1857, was the first to put forward the theory that history depended on the importance of food, soil and the general aspect of nature upon the formation of society. Adding 'the advance of European civilization is characterized by a diminishing influence of physical laws and an increasing influence of mental laws' and 'the measure of civilization is the triumph of mind over external agents.' Ten years previously Karl Marx had formulated the 'materialist conception of history.' Concluding that the organization of a society is conditioned by the economic circumstances of its existence, he laid down the principle that social relationships largely depend upon the particular dominant mode of production in any specific era and, therefore, that the principles, ideas and categories which are dominant in that era are no more permanent than the relations they express but, by contrast, are historical and transitory products.[2]

Between Marx's concept and that of the many historians whose treatment emphasizes the importance of kings, politicians, the landed aristocrats and the industrialists of the late eighteenth century, lies the socio-economic view of historical research taken in this study.

1.3 THE CONCEPT OF PROPERTY

Land, and the buildings erected on it, differ from all other forms of property since each square foot of it is unique, immovable and visible. In an urban setting in particular, it has the ability both to attract and repel social needs.

The existence of a shop attracts the traffic not only of its customers but the suppliers of the goods sold in it. It requires the construction of roads, drainage, water, gas, electricity and a host of other modern provisions both local and national. Economically, it is an extension of the factories that manufacture the very goods sold in it, the transporting of these goods and the monetary system that enables the proprietor, the customer and the manufacturer to operate. These externalities, important as they are, can be applied to other human activity, but property, unlike a motor car for instance, once constructed has a visible permanence and a place in the social order that gives it special significance. A growing awareness of these factors is shown by the numbers of national and local preservation societies and the measures taken by central government to hold public enquiries where new development of a significant nature is projected.

Though both economists are to the political right, neither Adam Smith writing in his *Inquiry into the Nature and Causes of The Wealth of Nations* (1766)[3] or Professor Hayek[4] in modern times, propose that property ownership or development should be of such an individualistic nature that it should escape from the need to comply with a well-regulated social order. Metaphorically, property is a motor boat whose internal combustion engine is inadequate to prevent it from being deflected off course by the forces of wind and tide. Whereas at the time of the Industrial Revolution, the property owner sailed in the calm sea of *laissez-faire*, subsequent tides of discontent fanned by the winds of social reform presaged a change in weather that has not yet abated.

The increase and relocation of population, the key ideas of clearly defined epochs and the effect of the legislation that ensued, take on a well defined pattern. Fiscal as well as planning control are apparent. Planning and use control followed new administrative structures and subsequently led, potentially at least, to virtual appropriation at last of profitability with the Land Commission and Community Land Act both no longer on the Statutes Book. The reaction of the private sector to these changes as well as the effect of exchange controls, excursions abroad and investment by foreigners in this country (affected both by inflation and fluctuations in interest rates) all play a part in the continuing saga of property development.

Whilst the specific issues mentioned above would seem to be the most convenient form in which to examine the pattern of forces that has resulted in the present state of the property market, the study relies to a large extent on the continuing effects, physical and moral, of the legacy that resulted from the problems created by the on-rush of the Industrial Revolution. This brought with it undisciplined urban expansion resulting in slums, disease and high mortality, especially amongst children. Those who at that time were responsible for providing both housing and places of work, failed not only in foreseeing the future (a failing that might be forgiven) but they also compounded this fault by neglecting to accommodate even the needs of their times. Those who view with hindsight do so with 20/20 vision and it remains to conjecture how observers a hundred years hence will comment on today's five million or so housed below the present legal standards.

The second half of the nineteenth century, whilst showing some effort on the part of enlightened reformers to ameliorate conditions, dealt only with the tip of the iceberg, and the seeds were sown for new collective ideologies. Perhaps in these circumstances will be found the roots of the belief that the ownership of property in private hands had failed the nation and that through the electoral system a Government would emerge to replace private landlords with a State monopoly. Certainly from 1945 to the present time there has been legislation in planning, taxation and the landlord and tenant relationship to a degree that had never before been experienced.

It is the aim of this book to provide sufficient detail to justify the proposition that the provision of the buildings where people live, work and spend their leisure time has been for too long subject to excesses of political dogma and subject to confusion and self-contradictions because of the inter-play between public and private factions.

Commenting on proposals put before the Cabinet, D. N. Chester,[5] who in 1945, was in the Economic Section of the Cabinet Office wrote

> 'The State leaves the actual ownership of the land in private hands, but takes away the profit motive, the mainspring of private enterprise, by nationalizing the development value. If I were labelling the scheme, I would call it "bastard Tory Reform". . . We feel that the scheme should either go a little further in the way of socialization or ownership. Any in-between system is likely to get the worst of both worlds.'

1.4 EARLY PATTERNS OF LAND OWNERSHIP

Because so much of the change in the land and property scene has resulted from the long-standing struggle between the opposing ideologies of socialism and capitalism, an examination of the history of land tenure must begin a long way back. In this book the word property, otherwise known as real estate, means *immovable* property, notably land and buildings. In earliest known history *movable* objects were the only form of property. There is no need for a particular reference to recall the history of the Indian tribes of the North American continent, living in tents which they moved from place to place as they followed the tracks of herds of wild animals over vast distances. Title over the land they covered was held or denied equally to both the buffalo and its Indian hunter.

Where primitive societies first turned to agriculture and the retention of more or less permanent farmland, then the community established ownership by right of possession. This right persists to this day in English Law as possessory title. The creation of a legal system over many centuries has (one hopes) produced a conveyancing procedure which is an improvement on the method of transferring property in olden times. This was the simple process of acquiring more land by eliminating neighbours with spear and sword. Certainly in England, and probably elsewhere, there developed a mutual protection arrangement where, either by amalgamation with adjoining communities, or through conquest, the smaller communities merged into larger ones, ruled by the physically strongest individual. This leader maintained his position through the fear of his subjects, though later he was chosen or elected on the basis of some other qualities. Thus, communities tended to grow and spread over increasing areas of land which in some cases led to the development of the idea of hereditary kings and kingdoms.

Pre-Roman history shows the division of England into numerous kingdoms where title to land was entirely possessory. This continued after the departure of the Roman legions and even later invasions from northern Europe barely altered the pattern of ownership. Perhaps the first positive change of this pattern came with the appearance in May 669 of Theodore of Tarsus[6] (602 - 690), 7th Archbishop of Canterbury. Bede declared he was the first archbishop to whom all the Church of the Angles submitted. The Saxon Kings, themselves already converted to Christianity, permitted the creation of regional (i.e. static) bishops subject to Canterbury. Previously, bishops had been appointed but they were merely permitted to roam the country preaching wherever they could find an audience. Each bishopric was now divided into parishes, often identical to townships. The King's representatives, the *Thegns* or local magnates gave, in the King's name, lands and emoluments and permitted the collection of ecclesiastical dues from the worshippers; thus Church Law, with its ultimate threat of excommunication and consequent damnation after death was able to create the first profit sharing system — tithes.

With the coming of William at the beginning of the eleventh century, virtually the whole of the land was conveyed to him by conquest and he, in turn, parcelled out his estate, giving large tracts to those of his followers who were later recorded in the Domesday Book as freeholders of their lands. Their tenure was secure as long as they gave fealty to the King and this was the 'simple fee' they paid for the grant. As late as the eighteenth century, deeds of conveyance often described land as being held in 'fee simple.' In their turn the Norman adventure knights turned land owners and were free to dispose of the whole or part of their estates, either by parting with the freeholds or granting lesser interests, which became known as leaseholds, copyholds and a variety of other holdings.

The pattern of land ownership was straightforward in that the freeholds were vested in the King's person, his knights and the Established Church. The populace lived on the land as serfs in varying degrees of thraldom, and were allowed to remain in their huts at the whim of their immediate lord or his representative on payment of military service or a proportion of their production. These grim conditions were, of course, lightened by the gradual erosion of feudal rights. Serfdom over the ensuing centuries gave place to small freeholdings and leaseholdings as the heirs to the landlords dissipated their inheritances, and gradually there came into existence the larger property owning community.

The gradual wresting of absolute monarchial power by the barons from King John, the formation of a Parliament and the institution of Common and Statute Law all played their part in promoting the interests of the individual's right to own property with no other threat to his tenure other than its appropriation should he commit a treasonable act.

1.5 THE EFFECT OF THE INDUSTRIAL REVOLUTION

Whilst agriculture and animal husbandry were the principal bases of the economy for the period before the eighteenth century and the beginnings of the Industrial Revolution, it was not surprising that there appeared in every decade individuals, who, with greater business acumen or by political means or even by close association with the fount of honours and privilege, the Crown, acquired large tracts of land and property. When the Industrial Revolution did come, it was often on these estates that were found the mineral wealth, the town extensions, the factory sites, and the lands through which canals and railways had to pass. Thus as the economy industrialized, fortunes were made not only in manufacturing but also in realizing the value of the mineral rights and the development value in land.

1.6 PHILOSOPHY, ECONOMICS AND POLITICS

In examining the events dating from the Industrial Revolution, it is enlightening to examine what attitudes have been taken by leading thinkers of earlier epochs. Almost all the most celebrated philosophers have commented on the part that the ownership of land plays in society and their views may be shown to have particular significance in the development of modern thought and the subsequent legislation. Among the earliest, the influence of Plato and his pupil Aristotle can be seen in the present day philosophies of the two main parties. Since the political parties have extensively expressed their opposing views on property in the form of actual legislation, it is important to trace their origins.

In Plato's *Republic*[7] he recommends that landed property be distributed in equal proportions among all citizens of a state. This proposition was actually earlier proposed by Phaleas of Chalcedon, to whom Plato refers. In the *Laws*[8] he holds that accumulation should extend to no more than five times the amount owned by any other citizen. Aristotle's *Politics*[9], in his chapter on Phaleas, refers to Solon who 'introduced laws restricting the amount of land which an individual might possess.' Other laws forbade the sale of property. No Locian could sell his property unless he could show beyond all shadow of a doubt that he had suffered some grave misfortune. Original allotments of land had to be kept intact.

Hippodamus, the first town-planner, albeit regarded as an eccentric, wrote of a city which he had in mind was to consist of '10,000 souls divided into three classes — one of artisans, one of farmers and a third of professional soldiers for defence.' He advocated dividing the land similarly into three parts — sacred, public and private. 'The first of these was to maintain the established worship of gods, the second to support the military, while the third was to be the property of the agricultural class.'[10]

Plato proposed 'thorough going communism': no private property beyond what was absolutely necessary, all were to live in small houses, eat simple food and have neither gold nor silver. Socrates extended this idea by proposing that all families should share houses and meals. Plato's *Republic* must be examined bearing in mind the times and conditions in which he wrote. He commented on a society that was mainly agricultural and not commercial. Whilst the patrimony of each household was to be strictly inalienable, the difference in property, both landed and personal, was to be kept within bounds by taxation, even to the extent of 100%.

Aristotle held somewhat differing views. 'How vastly more pleasant it is to be able to regard something as one's own.'[11] After commenting on the state of affairs in Sparta and Crete he maintained 'We cannot afford to disregard the experience of ages; you may be sure that if the methods advocated in Plato's *Republic* were sound they would not have gone unrecognised by so many generations.'[12] Aristotle wrote this less as commentary than as a refutation of the precepts of his erstwhile master, whose academy he attended from 367 - 238 BC. In his Book VI[13] (a postscript to Book IV) he discussed democracy in 'its truest form.' 'Advocates of democracy might well believe their constitution to embody freedom and equality.' From this base, he enquired 'which is the juster form of government . . . one based on property or one based on numbers?' His answer is clear: 'It is expedient as well as customary . . . that the most important offices are filled by election and open exclusively to those who have property qualifications.'

Namely to satisfy both the notables and the masses

1. No magistracy should be used as a means of amassing wealth.
2. While all citizens would be qualified for office the notables would actually hold office.
3. Office should be divorced from profit because the poor would rather devote themselves to private affairs since office carried no reward.
4. The poor would devote themselves to private affairs and so have the opportunity of becoming rich and the rich would avoid being governed by persons of inferior standing.
5. Inheritance should pass by descent and not bequest and no man should receive more than one inheritance.

On these points the ethical question is posed of whether 'a community should exist where the best things belong to a few with the majority having to be content with less - often much less.' Plato joins with Aristotle in accepting this. Much later we find Nietzche in agreement.

Comment by Aristotle on taxation is of some interest:

> 'extreme democracies are generally associated with large populations the members of which find it hard to attend the popular assembly unless they are paid for doing so . . . where there are no special revenues for this purpose a heavy burden falls on the notables from where the money has to be squeezed by means of property taxes, confiscations and corrupt tribunals all or which . . . have led to the overthrow of democracies.'

In the light of recent experience of almost non-stop legislation affecting the property industry of Great Britain it is impossible to resist the temptation to quote 'in the absence of such revenues therefore, the assembly should not meet too often.'

Solon (638 - 558 BC) is reported by Aristotle (who defends his constitution[14]) as having divided the population into four classes, the magistrates, archons and treasurers, superintendents of the State's prisons and clerks of the exchequer 'in accordance with the value of their RATEABLE property.' The Aristotelian view is of great importance in the history of philosophy since it is generally recognized as the main plank of philosophical thought through the centuries till the German Reformation when 'the mouth of Luther denounced the Aristotelianism of the schoolmen' but has since 'in the later nineteenth century . . . begun to be understood in terms of his own doctrine of evolution.'[15]

Plato's concept of communal living annoyed Aristotle. In his opinion it would lead to anger against lazy people. Property should be private but people who owned it should be trained in benevolence. Benevolence is a virtue but without property (this in its widest sense) would be impossible. 'Revolutions turn on the regulation of property' is rejected as a truism on the basis that the greatest crimes are due to excess rather than want; 'no man becomes a tyrant in order to avoid feeling cold.'

Antisthenes a disciple of Socrates, twenty years older than Plato, and who founded the School of Cynics, is said to have abandoned a life of ease in order to live as a working man and to preach to the uneducated. He advocated a return to nature and the abolition of government, private property, marriage and established religion. His disciple Diogenes had those and additional ideas. He wished to deface coinage. Coinage to him included 'every conventional stamp.' He considered false 'men stamped as Kings and generals, things stamped as honour, wisdom, happiness and riches.' All were 'base metal with lying superscription.'[16] Plato, Anthisthenes and Diogenes were but three among many who shared similar views varying between communal ownership of the land and communism in its modern form.

In comparatively more recent history (1372) Wycliffe[17], who incidentally might well be described as a Platonist, departed from his position of orthodoxy by advancing the theory that 'righteousness alone gives the title to dominion and property' and to some degree pre-empted Karl Marx by teaching that 'property is the result of sin.' Marx, of course, eschewed the word sin in favour of 'crime.' Wycliffe pointed out that Christ and the Apostles had no property and, therefore, the clergy ought to have none. These doctrines offended all the clerics (who owned property) but not the friars (who had none). Moreover, the Government of the day was happy to support his assertions for the Pope drew a huge tribute from England. It should be added that the Pope was subservient to France and England was at war with that country. Conversely, and a century and a half later (1518) Sir Thomas More's *Utopia*, intended no doubt as a political polemic in the Plato tradition, argued that the public good cannot flourish where there is private property and that 'without communism there can be no equality.'

Hobbes (1588 - 1679) studied scholastic logic and the philosophy of Aristotle, going up to Oxford at the age of fifteen. He held that the laws of property are to be entirely subject to the sovereign; 'for in a state of nature there is no property and, therefore, property is created by Government which may control its creation as it pleases.' His major philosophical work, *The Leviathan*, (1651) was preceded by the lesser known *De Cive* written in 1641 but not published until 1647 and which contains most of the political opinions, including that of property and the evils of democracy, that reappeared in *The Leviathan.* Since his main occupation from 1646 - 48 was to teach mathematics to the future Charles II, his views on property may be considered as pragmatic as those he held on censorship, 'a doctrine repugnant to peace cannot be true and, therefore, must be censored!'

If we are to rely on Bertrand Russell's interpretation on the rise of liberalism and its effect upon 'subsequent political and social developments'[18] we would accept that it came as a result of the Renaissance. It let to a rejection of medieval philosophy and politics as well as the resurgence of art and literature and promoted the belief that whilst all men are born equal their subsequent inequality is a product of circumstances. The 'divine right of kings' was rejected in favour of the view that every community has the right to choose its own government, a matter of particular concern to merchants whose activities were little understood by kings or aristocracies. The mercantile classes hoped to curb their previous masters, especially the Church whose opposition to scientific research was indirectly affecting commerce. Liberalism expresses optimism coupled with energy likely to benefit mankind materially rather than spiritually and thus it opposed both the Established Church and the modern Calvinists and Anabaptists. It turned its back on theological strife and politics and directed its energies towards science, the arts and commerce.

From the time of Alexander, individualism was preached by the schools of the Cynics and Stoics, the latter maintaining that a man could live a good life whatever his circumstances. This philosophy was continued by the Church before it controlled the State. Thereafter the Christian ethic taught through the Catholic Church that what was good was not to be determined except by the authority of that institution. Protestantism breached this custom when men came to believe that the General Council could err in matters of truth and that what was previously the monopoly of the Church was now fragmented into pieces to be picked up by any individual. This was epitomized by Descartes 'I think, therefore, I am.' It must be claimed, with some certainty, that the new philosophy, the mainplank of Protestantism, marked the departure not merely from previously accepted religious authority but from the social order that was controlled by it, and, furthermore, then opened for reassessment that order of which property formed part.

To quote Bertrand Russell 'individualism was idealistic in intellectual matters and economics but was not emotionally or ethically self-assertive.'[19] The earlier form of Liberalism was found in Holland and England, was mainly Protestant, valued commerce and industry and was by definition the philosophy of the rising middle classes rather than the monarchy and its attendant aristocracy. Above all, it had tremendous respect for the rights of property — rather than the divine right of kings — and especially property accumulated by the labours of the individual.

The great philosopher of the age, Locke,[20] influenced as he was by Descartes, first came into prominence with his *Essay Concerning Human Understanding* and quotations from this are of some importance. 'Where there is no property there is no Justice.' He distinguished between political and economic power and advocated that whilst a man should be allowed to leave his property to his children, political power should not so pass. Property figures prominently in Locke's writings and he regarded it as the chief reason for the institution of civil government. 'The great and chief end of men uniting into Commonwealth and putting themselves under government is the preservation of their property, to which in the state of nature there are many things wanting.' So convinced of this basic concept was he that he implied that those who have no property are not to be regarded as citizens and furthermore, 'the supreme power cannot take from any man any part of his property without consent.'

The implication that 'lackland' should not be regarded as a citizen ties neatly in with the fate of those who were denied the vote until the Reform Acts of 1867 and 1884. Women were still excluded, those over thirty years of age not obtaining the vote until 1918 and full equality with men taking another ten years. It is important to remember that property ownership until late in the nineteenth century was a pre-requisite to participation in the political life of the nation.

If property ownership was the qualification for the smallest participation in politics it carried with it obligations. Aristotle's pithy observations 'Men are easily spoilt: not everyone can wear prosperity' and 'Aim above all, by means of legislation, to prevent any man from becoming too influential either through his friends or through his wealth' are precepts too general to apply merely to property, but he went on to suggest rules that formed the background of behaviour of the landed aristocracy of England to well into the nineteenth century. The Stoics, Christians and later democratic factions of all types disagreed. The Stoics and the Church concentrated on spiritual values and the Christian ethic taught man to be virtuous in all circumstances. Democrats more concerned with politics saw property and the power it brought as part of the social ethos that somehow had to be 'just'.

One of the most profound influences on scholastic casuistry was the distinction made by Aristotle between wealth made by trade as opposed to that made by the skilful management of houses and land.[21] He particularly emphasized that true wealth is not coin and he condemned the use of coin for usury 'which makes a gain out of money itself and not from the natural object of it. For money was intended to be used for exchange but not to increase at interest.' Usury, now a term to describe moneylending at an exorbitant rate of interest, was from Bible times meant to describe moneylending at any rate of interest. From earlier times, however basic the economic group it was always divided into debtors and creditors. At most times landowners have been debtors and those in commerce, creditors. Each class had its own philosophy. Debtors naturally opposed usury whilst creditors equally approved of it.

The Church in medieval times was mainly supported by revenues from land and its disapproval of usury was evidenced in the rise of anti-Semitism, convenient when the traditional moneylender — the Jew — required the return of his capital lent perhaps to tide the borrower over a bad harvest. The Reformation, actively supported by the commercial classes, changed this thinking, since profits from trade could be employed in the lucrative business of lending at interest. The Calvinists and other Protestant sects approved, to be followed eventually by the Catholic Church — itself unable to resist the modern outlook. Philosophers whose universities derived their income from investment, and who ceased to be ecclesiastics, were similarly converted to the approval of lending at interest. Perhaps the Reformation might be considered as one of the reasons for the commencement of the system finally developed as today's money market.

If the route from Plato and Aristotle is to be shown to connect with their respective counterparts in today's thinking a sharp distinction needs to be made between land used for agriculture, the basis for most if not all the commentary till modern times and the expansion of the commercial and industrial use that marked the beginning of the Industrial Revolution.

Locke, who was not altogether the champion of the aristocratic capitalists as he has often been depicted, revealed in his writings a tendency to certain Socialistic principles. He suggests, for instance, that a man should own as much land as he could till. This was at a time when throughout Europe the bulk of agricultural land was owned by aristocrats who either exacted a rent, which could be varied, or a proportion of the lands's produce, sometimes as much as half. In Russia the proportion was determined by the minimum level of survival of the serfs who had no rights.

Born in 1632, Locke did not commence serious writings until after he had reached the age of fifty-four, whilst he was exiled in Holland. His value as a contributor to that part of philosophy covering property, is all the more valuable for his experience not only in study but as a practical man of affairs, since from 1696—1700, he was a Commissioner of the Board of Trade. His 'Treatises on Government' were meant to vindicate the Convention Parliament and the English Revolution of 1688, which refuted the idea of absolute monarchy. Indeed he may be said to have framed the principles later embodied in the French Revolution and the American Declaration of Independence.

At this time, the state of the rural English landowner was somewhat different from that of the rest of Europe. There were the Commons (land where ownership was vested neither in private ownership nor in the King's name) which enabled the labourer to share with his neighbours the right to raise a goodly proportion of his daily needs. The Parliaments were controlled by the great landowners who, using their legislative powers, had enclosed these commons since the time of Henry VIII. Each Act of enclosure which enriched the local large landowner forced the labourers towards starvation and thus to emigrate to the towns where the advent of the Industrial Revolution absorbed them. This movement of labour, gradual though it may have been, caused by a push from the land at the same time as a pull to the towns, played a most significant part in the pattern of property development.

The length of time it took before virtually all common lands were abolished is explained by the need for a separate Act of Parliament for each operation. The very growth of industry brought with it an increased demand for agricultural produce to feed the towns and agricultural workers' wages improved, thus slowing down the migration from field to factory.

1.6.1 The labour theory of value

It was Locke who first propounded the labour theory of value, i.e. the doctrine that the value of a product depends upon the labour expended upon it. The origin of the theory has been attributed firstly to Ricardo and then Karl Marx, but it seems to have been suggested to Locke by a line of predecessors going right back to Aquinas. The labour theory of value has two separate implications, one is ethical, the other economic. Ethically it asserts that the market value of a product ought to be proportional to the labour expended on it or that, in fact, the labour regulates the price. The economic implication is one of quantum.

Locke put a figure of nine-tenths of a product on the value for the labour but on the remaining tenth he was silent. So far as agricultural land is concerned he specifically mentioned the American Continent, then occupied for the most part by Indians, and placed no value upon this because the Indians did not cultivate it. In so doing he failed to realize that where people were willing to cultivate land it acquired value even before cultivation actually commenced. Additionally the buffalo, fed by the land, were essential to human survival. Although he favoured peasant proprietorship, he nevertheless failed to consider the situation where land became of value due to mineral deposits which required large capital investment, machinery and often a considerable organized work force.

It is obvious that land used for commercial and manufacturing processes in particular cannot support the principle that a man has a right to the produce of his own labour. In modern conditions of state-owned manufacturing organizations such as the motor car industry, railways, etc., the proportion of total output can never be precisely calculated in terms of labour.

Locke preached the theory of labour value throughout his career, supported by his research into the beliefs held by the older philosophers, particularly in relation to the effects of usury. He joined Ricardo in opposing usury and was followed later by Marx, who was in total opposition to Capitalists. Perhaps the significant factor underlying this opposition can be traced to a form of class warfare. Each in turn represented a 'class' opposition against one or other of the classes which the philosopher regarded as predatory.

Rousseau had this to say on the subject of land:

> 'The origin of civil society and of the consequent social inequalities is to be found in private property. The first man who having enclosed a piece of land bethought himself of saying "this is mine" and found people simple enough to believe him, was the real founder of civil society.'[22]

Thomas Hodgkinson, an ex-naval officer, published a relatively unknown but nevertheless important Socialist rejoinder (1825) to James Mill's theory (1817). His book *'Labour Defended against the Claims of Capital'* argued that if, as Ricardo claimed, all value is conferred by it, then all the reward ought to go to Labour. Hodgkinson's book indeed secured a reply from James Mill: 'their notions of property look ugly ... they seem to think that it should not exist and that the existence of it is an evil to them.' Mill's later letter condemning Socialism quotes, 'these opinions if they were to spread would be the subversion of civilized society: worse than the overwhelming deluge of Huns and Tartars.'[23]

Here a distinction must be made between what is political and economic on the one hand, and the history of philosophy on the other. Socialism had no philosophy at this point in time. It acquired it at the hands of Karl Marx whose influence on subsequent events has been so profound as to prohibit comment. Nevertheless, it is important to state that the 'philosophical dress' (to quote Bertrand Russell) which Marx gave to Socialism can be made without reference to the dialectic. The appalling effect of the Industrial Revolution on the nineteenth century as chronicled by Engels and the Royal Commissions led Marx to the belief that the economic system was likely to move from free competition towards monopoly and that in a highly industrialized community, monopoly would further what he saw as the increasing degradation of the labouring class. It was a simple step to advocate the state ownership of capital and land. Marx advocated that to achieve the millennium it is necessary to have a profound reorganization of social and economic relationships. Such a reorganization, which most people would term a revolution, has of course occurred subsequently in Russia in 1917 and since that date in countries whose inhabitants make up about a third of the World's population.

Not all philosophical utterances are drawn from people normally regarded as philosophers. To quote from Boswell's *Life of Dr Johnson*:[24]

> BOSWELL 'But, to consider the state of our own country; — does not throwing a number of farms into one hand hurt population?' JOHNSON 'Why no, Sir; the same quantity of food being produced, will be consumed by the same number of mouths, though the people may be disposed of in different ways. We see, if corn be dear, and butchers' meat cheap, the farmers all apply themselves to the raising of corn, till it becomes plentiful and cheap, and then butchers' meat becomes dear; so that an equality is always preserved. No, Sir, let fanciful men do as they will, depend upon it, it is difficult to disturb the system of life.' BOSWELL 'But Sir, is it not a very bad thing for landlords to oppress their tenants by raising their rents?' JOHNSON 'Very bad. But Sir, it never can have any general influence: it may distress some individuals. For, consider this: landlords cannot do without tenants. Now tenants will not give more for land, than land is worth. If they make more of their money by keeping a shop, or any other way, they do it, and so oblige landlords to let land come back to a reasonable rent, in order that they may get tenants. Land, in England, is an article of commerce. A tenant who pays his landlord his rent, thinks himself no more obliged to

> him than you think yourself obliged to man in whose shop you buy a piece of goods. He knows the landlord does not let him have his land for less than he can get from others, in the same manner as the shopkeeper sells his goods. No shopkeeper sells a yard of ribband for six-pence when seven-pence is the current price.' BOSWELL 'But Sir, is it not better that tenants should be independent of landlords?' JOHNSON 'Why Sir, as there are many more tenants than landlords, perhaps strictly speaking, we should wish not. But if you please, you may let your lands cheap, and so get the value, part in money and part in homage. I should agree with you in that.' BOSWELL 'So Sir, you laugh at schemes of political improvement.' JOHNSON 'Why Sir, most schemes of political improvement are very laughable things.'

Boswell quotes Johnson pontificating on much the same theme:

> 'Providence has wisely ordered that the more numerous men are, the more difficult it is for them to agree in any thing, and so they are governed. There is no doubt, that if the poor should reason, "We'll be poor no longer, we'll make the rich take their turn," they could easily do it, were it not that they can't agree. So the common soldiers, though so much more numerous than their officers, are governed by them for the same reason.'

However, history has proved him wrong on at least one thought:

> 'Mankind have a strong attachment to the habitants to which they have been accustomed. You see the inhabitants of Norway do not with one consent quit it, and go to some part of America, where there is a mild climate, and where they may have the same produce from land, with the tenth part of the labour. No Sir; their affection for their old dwellings, and the terror of a general change, keep them at home. Thus, we see many of the finest spots in the world thinly inhabited, and many rugged spots well inhabited.'

Whereas, before Marx, wars leading to conquest were waged between nations and differing religious sects, his contribution was to encourage conflict between classes of people irrespective of nationality or religion. In its simplest terms classes mean Capitalists on the one hand and the wage earners on the other and in the first category property owners figure prominently. The very arguments of the conventional economists such as Ricardo and Mill were turned neatly to reinforce his doctrines.

As argued previously the original doctrines of Plato and Aristotle may be seen as parallel streams of differing philosophies that have travelled down the ages, Plato to the left of centre, Aristotle to the right. If we examine the Platonic stream we find it was of lesser influence on the minds of men, as the idea of the commune as a substitute for unrestricted property ownership in private hands was either unacceptable or impossible to achieve throughout the ages until well after the Industrial Revolution.

Yet with Robert Owen and his contemporaries Hodgkinson and Cobden, the first steps towards Marx were taken. The failure of a wheat or potato crop is no less serious than defeat in war and can be followed by a revolution in the life of a community. The history of Ireland has shown that the potato famine played a more dramatic part in that country's story than Cromwell's expeditions. Three important factors contributed to the English famine — for such it was — that inspired Robert Owen. Firstly, the gradual extinction of commons reduced the ability of agricultural labourers to supplement their low wages by grazing a few animals or cultivating crops. Secondly, the onset of new factories, although providing work for the labourers drifting increasingly to the towns, depleted the numbers working on the land and reduced the output of corn. Thirdly, as a result of this the price rose and to protect the landowners a tariff was imposed on imported corn. This, of course, resulted in a demand for increased wages by the factory operators, threatening the profits of the industrialists who generally were not landowners. Thus the agitation for the repeal of the Corn Laws could be said to be not so much an agitation of the labourers but a confrontation between the landed proprietors and the new class of industrialists. Nevertheless, it was the workers who claimed as their own victory the repeal of the Corn Laws.

Thus, the Industrial Revolution is the watershed that has caused a quickening of the current of the stream to the left and all subsequent events point to its continuance. If the analogy can be further indulged, the stream to the right ran at its strongest at the commencement and for some time after the Industrial Revolution and at so fast a pace as to overflow its bounds into quagmires of social disgrace that succeeding generations have not yet been able to drain. In the efforts to do so, those who benefited most from the profits of the new economy had first to expose publicly the faults and then to remedy them, hence the spate of welfare legislation that characterizes the mid-nineteenth century.

With the outbreak of the 1914—18 War came the first of the attempts to control rents and give security of tenure to many tenants in the 'working-class' sector. Since that time, a see-saw motion best describes the imposition of controls by the left, and decontrol by the right. Creeping into the philosophy derived from the efforts of the left to distribute wealth more equally has been the creation over the last fifty years of extensive house and flat building by local and county authorities. Council houses and flats are principally occupied by working-class tenants whose votes at elections are mostly for the left. Thus, it is politically expedient for the left to create votes through the provision of municipal housing. Recognizing this trend, the right counter by selling council property to the sitting tenants in the belief that home ownership creates capital in the hands of the erstwhile labour tenant. No doubt the policy of the right has had some influence, if recent election results are any guide.

When dealing later with the events of the past fifty years, comment will be made on the effects of taxation and inflation that had much to do with the major shift towards commercial rather than residential development in that period since no matter who carries it out it is, in a democracy, just another commercial undertaking. Any undertaking will decline if legislated into unprofitability until finally there will be a dangerous inbalance with demand overwhelmingly exceeding supply. There appears to be a fusion of philosophical thought firstly with that of economics and then with politics. Until the Industrial Revolution, property was identified principally as agricultural land and the economists based their theories on its produce. The emergence of the United Kingdom as a manufacturing nation produced the beginnings of socialism and the subsequent notion of redistribution of wealth that included the concept of state ownership of property or at least its participation in the profits of development.

The unbridled development of slum cities to accommodate the workers left the Governments of the day no alternative but to enact the laws first to ameliorate the situation and then to prevent repetition of the worst of the excesses. It was becoming apparent that not only did adverse conditions produce discontent but the health of large numbers of people was affected: medical examination preceding recruitment of soldiers in times of war revealed that too many town dwellers were unfit for service. The ensuing legislation leading to town planning marched alongside the growing demand for overall socialistic demands for better conditions, including the provision of better housing and working environment. The former had its roots in the Victorian philanthropists of whom Peabody is a notable example, but they were able to deal only with the tip of the iceberg, it was a short step to municipal housing.

The situation concerning commercial property has arisen to a very large extent by virtue of the opposing philosophies of the left and right and mainly due to the use of taxation as a weapon by the left. It is significant to observe the way in which the Labour party, when in power, either nationalizes or sets up semi-nationalized undertakings such as transport, steel, aircraft and others, to be followed by the Conservative party returning parts of these industries into private ownership when they have the power so to do. The housing industry, which over the last four or five decades was drifting slowly from a private enterprise to a state enterprise through the agency of the municipalities, is now being transferred back to private enterprise, firstly by the sale of the tenanted stock to individual tenants and secondly by the withdrawal of grants to municipalities, forcing them to retrench. A Conservative Government repealed the Community Land Act in 1979 in order to reverse the Labour party's intentions to ultimately control the development of land or at least to take the major profit element away from private enterprise

A clear picture emerges from the study of legislation, especially in the last thirty years. The Town and Country Planning Act of 1947 commenced with the Labour party's intention not only to plan for an ordered use of all the land and buildings throughout the country, but more importantly to appropriate to the state the profits due to improvements by taxation in various forms. The Conservative Government did not repeal the Act in its entirety, keeping the use controls, but reversing the financial controls. This broad statement is in essence a measure of philosophical thought as it affects the parties and a study of subsequent annual Finance Acts may be said to equally reflect these attitudes.

Thus, it becomes a matter almost of prophesy as to the future of the pattern of property holding, and its development is dependent upon the compromised attitudes which will be taken by political parties, assuming that no one party persists in power, either permanently or semi-permanently, for extreme periods. Hugh Rossi MP, the spokesman for the Conservative party (in opposition prior to 1979) gave a lecture to the Incorporated Society of Valuers and Auctioneers in 1978. In answer to the question of whether the Community Land Act would be repealed and the Rent Acts modified to allow for a free market in all residential properties, he replied to the following effect:

> 'Should the Conservative party be elected in the future (as did happen in 1979) it would retain a tax on profits arising out of development although modified, but nevertheless above 50%, for it would be political suicide for the Conservative party to so modify legislation as to set up an entirely free market in residential property.'

The Labour party, on the other hand, has shown its hand clearly enough through resolutions passed at Labour party conferences where continual pressure from the extreme left has been directed towards increasing control in favour of tenants and towards acquisition of property in all its forms by the State. The question which is posed is whether the Labour party will firstly move further to the left, and secondly, whether in that frame of mind it will regain power in the near future. If this happens there will be a period of reimposition of both state acquisition in one form or another and taxation to reduce the profitability and thereby the impetus for property to be purchased and/or developed by private enterprise. This, in its extreme form could lead to a nation of non-property owners, except for an individual's own residence or small business premises. Should a more moderate element of the Labour party retain its influence, it may well come closer to the compromise to which the Conservative party has already committed itself. Namely that of retaining control of the use of property but taxing to some extent profits made from the ownership and development of property.

An attempt has been made in this chapter to trace the beginnings of comment by early philosophers concerning land and property and their counterparts down to the present day. The list of commentators could be extended, but the commentaries consistently fall into two main streams, one to the political left, the other to the right. It remains to evaluate the effect of philosophical thought through the ages on these considerations. Starting with the premise that the left favours a redistribution of wealth which the right opposes, it seems that those who vote for the left will have a greater say in government since traditionally the left draws its support from the working class who exist in the greatest numbers. Governments in the last thirty years or so have tended to be more in the hands of the left and consequently legislation has been enacted to reduce the profits made by developers when the left is in power and with few exceptions reversed or amended when the right has had the opportunity to do so.

Whilst it is rash to don the robe of an oracle, the vacillations observable over the last three or four decades tend to presume that between these two philosophies lies the future for property. If one may coin a phrase, 'politics and property are indivisible.'

REFERENCES

1. Buckle, H.T. (1857) *History of Civilization in England.* J.W. Parker and Son, London.
2. Marx, K. (first published 1847) *Misére de la Philosophie.* Published in 1974 by Lawrence and Wishart, London.

3. Smith, A. (first published 1766) *Inquiry into the Nature and Causes of the Wealth of Nations.* Published in 1977 by Dent, London.
4. Hayek, F.A. (1972) *Verdict on Rent Control.* Institute of Economic Affairs, London.
5. Minute from D.N. Chester to the Lord President of Ministerial Sub-Committee Compensation and Betterment (45) 6, 27 October 1945.
6. Trevelyan, G.M. (1970) *Shortened History of England.* Penguin, Harmondsworth, p. 66.
7. Plato, The Republic, 1266 b.
8. Plato, The Laws, V744 e.
9. Aristotle, *Politics and the Athenian Constitution.*
10. Herman, C.F. (1841) *De Hippodamo Milesio.* Marburg.
11. Aristotle, *Politics and the Athenian Constitution*, 1263[b].
12. Aristotle, *ibid.* 1264[d].
13. Aristotle, *ibid.* 630g[d].
14. Aristotle, *ibid.* Chapters 5—16.
15. *Encyclopaedia Britiannica* (1949).
16. Aristotle, *Politics and the Athenian Constitution*, Chapter 11.
17. Wycliffe, J. (*ca* 1375) *De Civile Dominio.*
18. Russell, B. (1961) *History of Western Philosophy.* Allen and Unwin, London, p. 127.
19. Russell, B. (1961) *ibid.* p. 580.
20. Locke, J. (1894) *Essay Concerning Human Understanding.* Oxford, Chapter III, Section 18.
21. Aristotle, *Politics and the Athenian Constitution*, 1257[d].
22. Rousseau, J.J. (first published 1754) *Discourse on Unequality.* Published in 1973 by Dent, London.
23. Mill, J.S. (first published 1910) in *Letters* (ed. H. Elliot), 2 volumes. Published in 1963—73 by Routledge, London.
24. Boswell, J. (first published 1831) *The Life of Samuel Johnson.* Croker. Published in 1980 by Oxford University Press.

FURTHER READING

Briggs, A. (1963) *Victorian Cities.* Odhams.

Burke, G. (1971) *Towns in the Making.* Edward Arnold, London.

Burke, G. (1976) *Townscapes.* Pelican, Harmondsworth.

Chalklin, C.W. and Hariden, M.A. (eds) (1974) *Rural Change and Urban Growth 1500—1800 — Essays in English Regional History.* Longman, London.

Cheshire, G.C. (1976) *Modern Law of Real Property.* Butterworths, London.

Corfield, P.J. (1982) *The Impact of English Towns 1700–1800.* Oxford University Press, Oxford.
Geddes, P. (1968) *Cities in Evolution.* Ernest Benn, London.
Hoskins, W.G. (1960) *The Making of the English Landscape.* Hodder and Stoughton, London.
Jenkins, S. (1975) *Landlords to London.* Constable, London.
Lambert, C. and Weir, D. (eds) (1975) *Cities in Modern Britain.* Fontana, London.
Munford, L. (1961) *The City in History.* Secker and Warburg, London.
Sharpe, T. (1950) *English Panorama.* The Architectural Press, London.

2

From the Industrial Revolution to World War I

The Industrial Revolution not only changed the life-style of the majority of the population, it had a profound effect on the small caucus who wielded the power to rule the country. Hitherto, it was always thought that those who held the landed estates on which the economy was based controlled the 'gross national product' (a modern phrase). The Revolution brought into being a new national product leading eventually to the mass production of manufactured articles requiring a small area of land, and this resulted in the gradual shift of power to industrialists.

2.1 POPULATION TRENDS

Shifts in population soon resulted in the enlargement of towns and the partial depopulation of the countryside. Between the years 1801 and 1914 the population of England, Wales and Scotland rose from 10 501 000 to some 40 000 000. Obviously, as movement from the country to the towns continued, the children born to the re-settlers increased the trend. The population of Glasgow, which in 1801 contained just over 5% of the population of Scotland, more than doubled by 1851 and by 1891 the city contained 19.4% of that country's people. The continuance of this trend into the twentieth century is exemplified by Table 2.1 taken from the National Register, Statistics of Population. By 1851 there were more town dwellers than country dwellers in Britain

Table 2.1

Year	*Population*
1801	10501000
1811	11970000
1821	14092000
1831	16261000
1841	18534000
1851	20816000
1861	23129000
1871	26072000
1881	29710000
1891	33028000
1901	37000000
1911	40831000

2.2 NOT SO MERRIE ENGLAND

The inventiveness and energy of a manufacturer led to profits which gave him the wealth wherewith to buy the land necessary to carry out production, but here the pattern changed and could be regarded as a seizure of landed property and even movable property by reverting to acquisition by conquest. The weapon used by this new class of conquerors was the profit from manufacturing and trading.

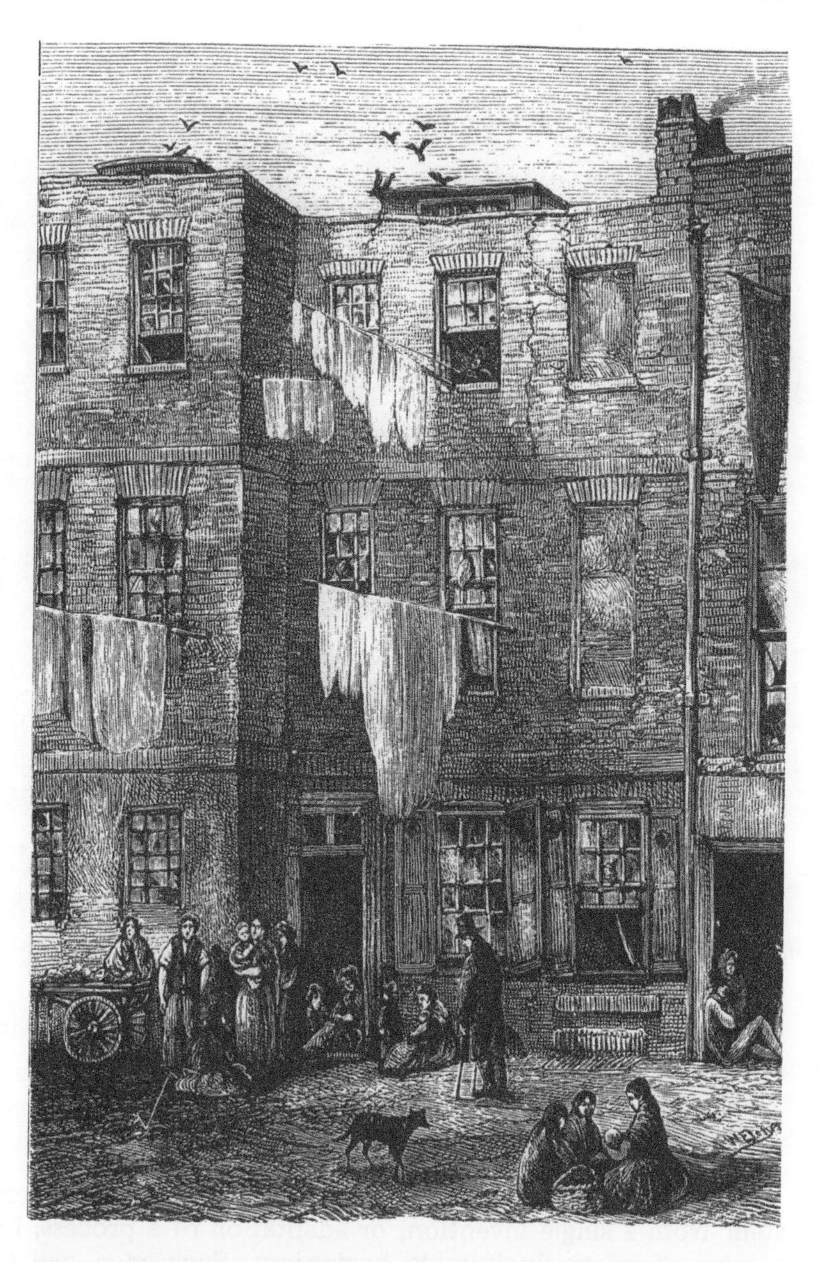

Plate 1 Wild Court, Seven Dials, London. A typical nineteenth century slum. (Courtesy of Mary Evans Picture Library.)

Plate 2 Devil's Acre, Westminster. Another example of nineteenth century slum housing. (Courtesy of Mary Evans Picture Library.)

The beginnings of almost every great industrial concern stem from a single invention, or adaptation of a process, by a man of relatively humble beginnings. Such men had to make, by self-sacrifice, the saving necessary to accumulate the capital needed to commence the small operation which could eventually grow into a large concern.

The very process formed character, and this took the

shape of making the most out of every penny invested. This process had severe social consequences for those employed as labourers:

> 'In one cul-de-sac, in the town of Leeds, there are thirty-four houses, and in ordinary times, there dwell in these houses 340 persons, or ten to every house; but as these houses are many of them receiving houses for itinerant labourers, during the periods of hay-time and harvest and the fairs, at least twice that number are then here congregated.
>
> The name of this place is the Boot and Shoe-yard, in Kirkgate, a location from whence the Commissioners removed, in the days of cholera, 75 cart-loads of manure, which had been untouched for years, and where there now exists a surface of human excrement of very considerable extent, to which these impure and unventilated dwellings are additionally exposed. This property is said to pay the best annual interest of any cottage property in the borough.'[1]

> 'Canning Town built in the 1850s to accommodate the adjacent dock workers took no account of the site in Plaistow Marshes below the level of the Thames high water where roads were not made up, where there was no proper drainage: in one street the only drain was beneath a pump and passed into a well which was the only source of drinking water.'[2]

One of the slums of Glasgow was the subject of a report which read:

> 'The houses in which they live are unfit even for sites, and every apartment is filled with promiscuous crowds of men, women and children, all in the most revolting state of filth and squalor. In many houses, there is scarcely any ventilation; dunghills lie in the vicinity of the dwellings; and from the extremely defective sewerage, filth of every kind constantly accumulates. In these horrid dens, the most abandoned characters of the city are collected, and from thence they nightly issue to disseminate disease, and to pour upon the town every species of crime and abomination.'

'It is not surprising that the death rate in Glasgow which was one in thirty-nine of the population in 1821 rose to one in thirty in the following year, one in twenty-nine in 1825, one in twenty-six in 1831, and that the mortality of children under the age of ten had risen from one in seventy-five in 1821 to one in forty-eight in 1839.'[3]

Further figures showing the difference in the average age of death in drained and undrained districts[4] led to recommendations by the Royal Commission of 1844–45, the Municipal Reform Act, the Nuisances Removal Act of 1846, the Towns Improvement Clauses Act of 1847; the Public Health Act of 1848, the Common Lodging Houses Act and the Labouring Classes' Lodging Houses Act of 1851.

In 1885, in a report by William Farr[5] the average infant mortality was stated to be 158 per 1000 births and was 111 per 1000 births for selected 'healthy districts'. In Liverpool, Leeds and Leicester it was over 200 in each case. He wrote:

'So unfavourable to infant life are the unsanitary conditions of large towns — expecially Liverpool — that not only is the mortality at some months of age twice as high as it is in the healthy districts, but at seven months of age and upwards it is three times as high. The mortality of infants by lung diseases in Liverpool is higher than in any other large town.'[6]

In terms of loss to the community by capitalizing actual income at 5% interest for each year of life, William Farr valued the population at £5250000000. He argued that had the unhealthy districts the same mean expectancy of life as the healthy ones, this figure would have gone up to £6300000000.[7]

2.3 THE BEGINNINGS OF REFORM

Edwin Chadwick, probably the greatest sanitary reformer of the age, laid down the following criteria for judging the sanitary conditions of a town.

'He would have the sanitary doctor inquire into the state of the intestine or sewerage of the place. Are the great

Plate 3 Belgrave Square, London. A typical example of private enterprise town planning in the nineteenth century. (Courtesy of Mary Evans Picture Library.)

Plate 4 Berkeley Square, London, *ca* 1860. Another instance of private enterprise town planning. (Courtesy of Mary Evans Picture Library.)

canals properly purged and cleansed? Is the breath of the place sweet and wholesome? Is it free, or is it infested with vermin? Is the circulation of what goes in and out of the town orderly and regular? Is the water with which it is supplied of good and proper quality? Is the food sufficient in regard to quality and quantity? Is the place supplied with pure air or does mist hang over it morning and evening like a fog? Is the mental condition of the place good? Is it free of discontent, irritation, or excitement? Is the death-rate that of a healthy community, and is the hereditary history of the town of such a character as to be creditable to its constitutional qualities? In a word, is it a town that an insurance company could insure wholesale without weighting it with any excess on the normal premium? If the answer to all these questions be the affirmative, then the town may be pronounced healthy. If it fails to give so clean a record, then the sanitary doctor is to prescribe for it sanitarily, as the curative doctor might, in his way, prescribe for a sick man.'[8]

2.4 PRELUDE TO PLANNING

The conditions under which so many people were living were the result of employees housed in accommodation built by industrialists who aimed to make the most intensive use of the land at the minimum cost. This produced a high income and had little or no hindrance from the law as it then stood. The tenants were not sufficiently organized to speak with one articulate voice, although it is true that there were those who recognized their plight and in *The Builder* the editor, George Godwin, wrote 'Running alongside the Acts of Parliament were the more practical efforts of those who set about producing a better standard of accommodation for the working-classes.' Unlike present day municipal housing schemes, these pioneers somehow succeeded in keeping rents down to a point that the tenants could afford, and showing a return on capital (since interest rates ran at 4–5%) that certainly made it worthwhile as a straightforward business proposition. Amongst the earliest was the Metropolitan Association for Improving the Dwellings of the Industrious Classes.

Built in 1846, the first of its developments cost £43 3s 4d, a room, and by 1870 the return on invested capital was 5¼%.[9] Others concerned in what came to be known as 'artisan dwellings' were the Improved Industrial Dwellings Company, (IIDC), and the Peabody Trust. The IIDC's construction costs came out at £40 per room and showed an investment return of 8%.[10] It later borrowed money from the Government at 4% and covenanted to pay no more than 5% (tax-free) by way of dividend. The Peabody Trust, although subsidized by its founder to the tune of £500000 showed only 2% at an average cost of £113 per room.

Octavia Hill remarked in 1875 that private efforts had housed only 26000 people in the previous thirty years.[11] She also pointed out that London's population increased by that amount every six months. By 1905 there were nine principal associations in London which housed no more than 123000 people. In the provinces there was precious little work done. There was an overwhelming need for artisan housing, and it was possible to meet this need with a profit to the developers, yet the totality of their efforts proved in the last resort, inadequate. The same story can be said for the housing societies, and the building society movement had played no material part in the provision of modern accommodation at that time.

It is significant that so many scars of the Industrial Revolution remained at the turn of the twentieth century. In spite of efforts by successive Parliaments it seemed that the total effect of legislation was merely to prevent a repetition of what had occurred before, and even to this extent, what was achieved proved ineffective, since the design of much of what was rebuilt fell below the standards which the developers might well have imagined would be acceptable to future generations. Lord Shaftesbury's Act of 1851,[12] whilst giving local authorities powers to build working-class houses, did not in the end prove to have added much to the stock and most of the worst property remained. It was left to local authorities through their various departments, including the Commissioners of Police, to promote private Acts of Parliament.

An interesting sidelight is revealed in the 'Report on Housing and Industrial Conditions and Medical Inspection of School Children' commissioned by the Dundee Social Union in 1905. In 1905, the Prudential Assurance Society informed the Union it would not issue life policies in certain districts without special enquiries, in view of the sanitary conditions.

The nineteenth century saw a change in the building industry. Large builders, especially those operating in the expensive residential areas, were able to plan estates as a whole, but the mass of cheaper building was done by small firms and individual builders constructing a few houses at a time. Gradually, the master carpenters, bricklayers and masons disappeared to be replaced by the larger contractors.[13–15] Regulations laid down by the Public Health Acts of 1875, 1890[16] and 1891[17] (which applied to London only) specified higher standards but building byelaws nevertheless were difficult to enforce, and it was not compulsory for local authorities to initiate them. District surveyors were liable to pay costs out of their own pockets if actions against infringers of the Building Acts were unsuccessful.[18] The costs involved were paramount in the mind of local authorities who might have otherwise carried out improvement schemes for improving sanitary conditions, or building more houses. In 1881, the Chairman of the Commissioners of Sewers of the City of London commented sadly on two improvement schemes carried out by his authority: 'If we had given every man, woman and child £100 or £150 to start them in life somewhere else, it would have been cheaper to the ratepayers.'[19]

Fortunately, financial considerations working in reverse cleared considerable areas of slum property. Especially steep rises occurred in the prices of land suitable for commercial property in the Manchester area before 1900. In the neighbourhood of the Bank, it was believed to be approaching £1 million per acre in value[20] and in Piccadilly and Corporation Street, the land values trebled between 1862 and 1871. In Market Street and Cross Street, the value had risen from between £10 and £25 to £85 and £95 per square foot. Not all local authorities were concerned with profiting from redevelopment. As they were obliged to rebuild working-class dwellings, the Metropolitan Board of Works in carrying out an improvement scheme in Whitechapel and Limehouse sold to the Peabody Trustees for £10000 a commercial site for which it could have obtained £54000. Thus, the rate payers had to foot the difference of £44000.[21]

2.5 EARLY ATTEMPTS TO TAX DEVELOPMENT

About this time, the power of compulsory purchase given to local authorities raised the question of betterment, this came about as compensation for compulsory acquisition and was based on existing use value.

Invariably as development took place, adjoining property rose in value. In London, seven houses in Cannon Street valued at £2770 in 1871 were, four years later after the construction of Queen Victoria Street, assessed at £3794. The original concepts of recouping betterment might well be contained in the Statute of 1427, when the Commissioners of Sewers were empowered to collect a levy on works for sea defence.[22] Subsequently, provision was made in 1662 for a tax on values increasing as a result of street widening due to the rebuilding of London after the Great Fire, and similar provisions are seen again in the Edinburgh Improvement Act of 1827, supplemented by an Act of 1832[23,24], but these lapsed in 1851. Betterment in a restricted form again appeared in the amendment to the Torrens Act in 1879, which provided for compensation to be reduced by the amount of betterment arising from the execution of the same scheme. It was left until 1895 for a direct betterment charge in connection with the Tower Bridge Southern Approach and the first General Act for direct levy of this charge was provided in the 1909 Housing and Town Planning Act.

Although in 1894 the London Building Act laid down new principles, including the rights of light and air, statutory and bye-law controls of new building had many deficiencies and these made little difference to the appearance and condition of towns up and down the country. With a growing public awareness of minimum conditions imposed and watched over by the local authorities, street after street of identical and depressing slum properties disappeared, although these were only to be replaced by an improved version of the same, conforming merely to new sanitary conditions including proper air and light. The local authorities had been given powers under Lord Shaftesbury's Act of 1851[12] to build working-class houses, but there is little evidence that much was done, and although encouraged by private Acts of Parliament, some local authorities did no more than demolish unhealthy areas without the rebuilding permitted under the same Act.

It would seem that the nineteenth century closed with much of the problem of providing satisfactory housing for the working classes unsolved, although there was a growing clamour amongst the occupiers. This was realized by politicians and by a strong lobby of enlightened industrialists, some posing as philanthropists, others genuinely moved by charitable motives. Looked at in retrospect, all that was achieved in the closing years of the century was to prevent the worst from happening again without providing any positive plans for significant improvements in the future. Many of the Acts of Parliament were permissive rather than mandatory and it can be concluded that the twentieth century dawned with no positive planning.

Although comment is made on early efforts of voluntary planning in examining the effects of planning law on the development of property, the gradual change from a more permissive to a rigid mandatory legislation produced results tending to direct developers' efforts into clearly defined channels of use, density and location. Additionally, the developer was threatened as early as 1911 with specialist taxation, to be enacted in subsequent planning legislation. Those who wish property development profits and land fit for development to be in public rather than private ownership have traditionally regarded planning legislation as the first step in the march towards the nationalization of a significant proportion of property in the United Kingdom. In the general sense, though, good planning need not be dependent on legislation and self-imposed aesthetic and financial disciplines have produced socially acceptable results, albeit on a limited scale.

Various planning measures and their repeal and subsequent re-imposition by successive governments in line with political aims have left the public, the planners and not least the developers dissatisfied with the prevailing system which invites comment into its origins, aims and results to date. The story commences with the General Election of 1906, in which a substantial majority was gained by the Liberal party, whose election platform promised improvements in housing and the quality of the environment. Underlying these aims was the urgent problem of recruitment for the army. In 1871, Dr H. W. Rumsey[25] published his findings on the proportion of rejection amongst army recruits, which he thought was related to poor living conditions in towns and cities. In 1904, the Committee on Physical Deterioration reported that although evidence was not conclusive, there was, nevertheless, a distinct deterioration of physical standards in the working classes, and it was concluded that overcrowding and the pollution of the atmosphere were the cause of a generally low standard of physical fitness. With regard to town planning conditions which had already proved injurious to health were being repeated. The very real problem that was being stated was the difficulty the army had in finding suitably fit recruits for the Boer War. The report went on to say:

> 'In England no intelligent anticipation of a town's growth is allowed to dictate municipal policy in regard to the extension of borough boundaries, with the result that when these are extended, the areas taken in have already been covered with the normal type of cheap and squalid dwelling houses, which rapidly reproduce on the outskirts of a city the slum characteristics which are the despair of the civic reformer in its heart. . . In this connection it would be expedient to secure the co-operation of Local Authorities in contiguous areas that are becoming rapidly urbanized.'

Slums essentially occurred for a number of reasons, but chiefly due to the cupidity of late eighteenth century and early nineteenth century factory owners who, being forced to provide accommodation for their own employees, met the demand at the lowest possible expense. Slums also appeared in areas where the occupants were not employed by the landlords, notably in the centres of towns established well before the eighteenth century. It is likely that houses built before that time had been allowed to deteriorate and that population increases leading to increased demand had gradually encouraged landlords to allow multiple tenancies where previously one family had occupied a single dwelling. The combined rents of floors or rooms let singly inevitably would have exceeded the income derivable from the house let to one family and it is reasonable to suppose that excessive use plus the non-caring attitude that goes with multiple letting produced the conditions which led to the creation of a slum. In particular, sanitary arrangements of a primitive nature that might have sufficed, under single family control caused health hazards when shared with many others.

There was even more to come on this theme. In 1901 T. C. Horsfall wrote:

> 'Unless we at once begin at least to protect the health of our people by making the towns in which most of them now live, more wholesome for body and mind, we may as well hand over our trade, colonies, our whole influence in the world, to Germany without undergoing all the trouble of a struggle in which we condemn ourselves beforehand to certain failure.'[26]

Then, as in the 1930s, there were those who were reading portents of a coming war with a nation that already had a well-organized system of roads and activity zones for future expansion in every town. The German system gave to their towns and cities power to buy land on the outskirts. In short, one rationale for the introduction of a Town Planning Act in Britain was to help ensure the future defence of the realm, following the Rumsey Report of 1871. The Defence of the Realm Act, which came some years later in the middle of the 1914–18 War, is now almost entirely remembered for the restrictions it imposed on the opening hours of licensed premises.

2.6 THE FIRST PLANNING ACT

The first part of the 1909 Act had as its objective:

> 'To ensure by means of schemes which may be prepared either by local authorities or landowners, that, in future, land in the vicinity of towns shall be developed in such a way as to secure proper sanitary conditions, amenity

> and convenience in connection with the laying out of the land itself and of any neighbouring land.'[27]

The essence of the second part of the Act was the permissive powers it gave to local authorities for the preparation of 'schemes' for controlling the development of new housing areas (Table 2.2). The interpretation of a scheme was given in the fourth Schedule, which listed the most important constituents of a scheme as the streets, roads and other thoroughfares, highways, building structures, open spaces, the preservation of historical buildings and objects, the preservation of beauty spots and, of course, the more practical ingredients of sewage disposal, lighting, water supply, etc.

Annual Reports from local authorities, covering a period of some eight years, show that no more than 199 schemes were drawn up, emphasizing the ineffectiveness of permissions compared with mandatory planning legislation. There were provisions for compulsory purchase, and significantly, the Act provided for the collection of a betterment tax at 50% on any increase in property values due to the operation of a scheme. At the same time as this Act was passing through its various stages, as a Bill, towards enactment, Lloyd George introduced a land tax as part of his budget. The Bill became an Act just one week before the Government fell defeated on the land tax clauses in the Finance Bill.

A number of new factors had been appearing which had some relevance to the timing of the 1909 Act. It could be said that development was at that time running along three main lines, namely, the reconstruction of the inner areas of cities and towns, suburban extensions to the same towns, and the building of garden cities. The clearance of insanitary houses and their replacement in many cases by commercial properties took place, for the most part, in the centre of cities where actual conditions made it imperative for the local authorities to take action. They were assisted by the rise in land values due to the demand to rebuild commercial premises on the cleared, previously residential sites.

Surburban development owed its impetus to a reduction in working hours. In 1886, it was estimated that since 1870 the working week had fallen by an average of three to four hours per week. Casual, non-skilled labour was giving place to skilled

Table 2.2

Year ended 31 March	*Urban districts*		*Rural districts*		*Total*	
	Number of schemes	*Total acreage*	*Number of schemes*	*Total acreage*	*Number of schemes*	*Total acreage*
1911	3	9668			3	9668
1912	11	4174			11	4174
1913	17	27563	2	9173	19	36736
1914	17	23060	6	16731	23	39791
1915	42	60211	7	16991	49	77202
1916	16	33249	2	7335	18	40584
1917	12	24493	3	9582	15	34075
1918	1	306			1	306
1919	12	27239	4	9409	16	36648
1919*	14	28259	3	11553	17	29712
Total	145	228122	27	80774	172	308896

*1 April to 30 July 1919

and non-manual labour, so that a greater proportion of city dwellers were leading more organized lives with shorter hours and, therefore, more time to travel and wages were increasing. The improvement in railway services made it possible for the Town Clerk of Shoreditch to report in 1902 that it was just as quick and cheap to travel by rail from Enfield to Shoreditch as it was by tram from one end of Shoreditch to the other. Development in the suburbs obviously scored over city centre locations on land prices. The impetus outwards was aided by factory developments which were advertised as 'trading estates.'

2.6.1 Examples of voluntary planning

In 1896, W. P. Hartley, the jam maker, set up in Aintree and Trafford Park Estates began a trading estate as a result of the opening of the Manchester Ship Canal and the Manchester Docks. Although these pioneers were not destined to be copied in any great number until after the First World War, the pattern of trading estate development had been set. These developments required the provision of both rail and road transport and this topic will be examined as a separate issue.

There were disadvantages of suburban life as pointed out by the Social Science Association.[28] These were faulty building, bad drainage and insufficient water supply caused by the lack of legislation over suburban development. Whereas in central areas planners could control new building through byelaws, suburban areas were not subject to similar controls. The Association recommended that both municipal areas and their suburbs should be brought under the control of one sanitary authority. The earliest of the suburban developments were described as 'jerry-built' and chaotic,[29] but time and changes in income, together with public demand, brought improvements and in 1877 the eminent architect Norman Shaw was commissioned to create Bedford Park (in south-east London). This was a great improvement over previous housing development and provided a high standard of design. New standards of layout were set by Port Sunlight and Bournville which, although not suburban developments, nevertheless served as models for what was to follow.

Yet another innovator was the Tenant Co-operators Limited,

a company which raised capital for residential developments from the public, many of whom were prospective occupiers. Interest paid to investors was limited to a maximum of 5%. Part of the profit was credited to the tenants in shares and the profit surplus (after interest was paid) was distributed as a dividend on rental, copying the 'divvy' paid to customers by the co-operative societies. Hampstead Garden Suburb was a supremely successful example of this system.

Perhaps of greater significance was the more adventurous type of development which came to be known as the 'garden city,' the modern successors of which can be found for example in Basildon New Town and Milton Keynes. Samuel Oldknow began the trend in 1790 by building around his spinning mill at Marple a 'contented' community, thus deliberately creating employment for heads of families. As this enterprise was so expensive, Oldknow relied on the employment of parish apprentices;[30] young persons taken from parish-maintained workhouses. The apprentices' 'wages' were paid direct to the parish, the latter often commuting these in lieu of one capital sum. Charles Dickens' *Oliver Twist* gives some idea of conditions suffered by these unfortunates, who could almost be considered as providing slave labour.

The social reformer Robert Owen built a factory and housing estate at New Lanark in 1784, transforming the village by increasing the population by over 1100 persons. Many other schemes of this nature were proposed, although few were actually started. In 1851 Titus Salt, at Saltaire near Shipley in Yorkshire, created 'a complete town with wide streets, spacious squares, with gardens attached, ground for recreation, a large dining hall and kitchens, baths and wash-houses, a covered market, schools, and a church; each combining every improvement that modern art and science have brought to light, are ordered to be proceeded with by the gentleman who has originated this undertaking.'[31] Salt's example was followed by Edward Akroyd at Copley near Halifax 'with allotment gardens, a recreation ground, a church and a village school, of which one classroom served also as a public library and newsroom.' George Cadbury built Bournville, and W. H. Lever Port Sunlight. All of these developments caught the imagination of large-scale employers such as the British Aluminium Company who built near Loch Ness and the Brodsworth Main Colliery Company, who built the Woodlands Colliery village near Doncaster.

Probably the last important example of this kind was that of Messrs Reckitt who created the Hull Garden Village Company developing no less than 140 acres. Sir James Reckitt's speech at the opening ceremony is worth quoting because it shows an awareness of circumstances which gave rise to future political thinking:

> 'The only object in view, is the betterment of our neighbours and to enable them to derive advantage from having fresh air, a better house, and better surroundings I . . . urge people of wealth and influence to make proper use of their property to avert *possibly a disastrous uprising*.'[32]

Here at least was one sensible property owner — albeit one whose main interest was the production of little blue bags used to whiten washing -- realizing that his work people had to be provided with decent housing before it was demanded by them. This might be said to be somewhat on a level with an apocryphal story told of Henry Ford, who it was said once crept into his workshop in the dead of night and added to the graffiti on the toilet walls by scribbling in large letters a demand for a wage increase of five cents an hour, urging a strike if not conceded. He reckoned that had he not stirred unrest at five cents, at least one of his rabble-rousing workmen would have asked for ten.

The idea of a garden city as distinct from suburbia or city life, was made popular by Ebenezer Howard's book, *Garden Cities of Tomorrow*[33] and a revised edition of his original book entitled *Tomorrow — A Peaceful Path to Real Reform* suggested garden cities of some 32 000 inhabitants at maximum. Should the need for expansion arise, then another garden city should take the overflow in a location some distance from the original, so that the end result would be a cluster of garden cities, interspersed with open country. Howard, a man of forceful personality, and enthusiasm, initiated the formation of the Garden City Association which, in turn, under the name of the Garden City Pioneer Company, developed Letchworth Garden City calling it First Garden City Limited. It was hampered by a shortage of capital, and although building commenced in 1903, no dividend at all was paid for ten years. Moreover, the maximum permitted level of interest of 5% was not reached until 1923, and it would seem that inflation alone saved the venture from being a financial loss, although it was undoubtedly a social success. The critics of Bourneville, Port Sunlight and Letchworth, saw them merely as a new kind of suburb, and observed that in private hands a viable financial success was difficult, if not impossible, more often than not depending on the philanthropy of individual manufacturers. Though they combined this with self interest they followed (one suspects) a little of the philosophy of Sir James Reckitt.

The huge amounts necessary to promote garden cities led inevitably to the idea of the hybrid 'suburb on garden city lines' the most successful of which was the Hampstead Garden Suburb. In July 1905 Henrietta Barnett's introduction read

> 'We desire to do something to meet the housing problem by putting within reach of working people the opportunity of taking a cottage with a garden within a 2d., fare of Central London, and at a moderate rent. Our aim is that the new suburb may be laid out as a whole on an orderly plan. We desire to promote a better understanding between the members of the classes who form our nation. Our object, therefore, is not merely to provide houses for the industrial classes. We propose that some of the beautiful sites round the Heath should be let to wealthy persons who can afford to pay a large sum for their land and to have extensive gardens. We aim at preserving natural beauty. Our object is so to lay out the ground that every tree may be kept, hedgerows duly considered, and the foreground of the distant view be preserved, not as open fields, yet as a gardened district, the buildings kept in harmony with the surroundings.'[34]

To enable this project to proceed, a Private Act of Parliament was necessary to overcome byelaws,[35] amongst other provisions enacted which differed from the general building laws were:

1. The average plot ratio, i.e. development density of houses, was limited to eight per acre.

2. Building lines were fixed at 25 feet from the centre of any road.
3. The Company was given power to regulate gardens and open spaces for common use.
4. The law against cul-de-sacs was relaxed.

Once again, whilst it provided a model for many developments that followed, the housing on that part of the estate reserved for the well-to-do inevitably progressed far more quickly than that at the other end of the scale. The company limited its dividends to 5%, but its history of payments was disastrous, and from 1914 until 1933 it suspended payments. Once more it was shown that such a venture, whilst successful in terms of reputation, failed in its social aim to provide a mix of classes and it was only inflation that rescued it in the end from becoming a financial failure.

More realistic perhaps were the efforts of private interests with no avowed pretensions of philanthropy or reform, but with the simple object of treating land and housing as commodities which could be the basis of trade in its existing or manufactured sense. They thought of the building of a house as the same as the manufacturing of the brick which was a part of it. Samuel Oldknow (referred to previously) and his like seemed to have grasped the nettle that bad housing, like bad food, was likely to engender disease and that profit motives alone were not sufficient if the housing conditions of the work force were likely to endanger business efficiency. The physical condition and the morale of the force, as Reckitt pointed out, could have a possible political reaction.

Much of the property developed at the height of the Industrial Revolution was the responsibility of the factory owners who rented out the houses to their workers and, in effect, placed themselves out of the competitive house sale or letting market. Their housing standards were so low that it is no wonder that the ensuing results caused social unrest leading to inevitable political action and consequential legislation. The market which developed on straightforward competitive lines produced no social outcry and the cottages of Chelsea originally occupied by the artisans of London remain as *bijou* residences for the middle and indeed wealthier classes. The developers of this latter type of property and those catering for the wealthy, laid out their housing estates with only the prospect of profit from sales and lettings and were doubtless influenced by customer demand to provide pleasant squares and streets and attractive exteriors (in the fashion of the time and with many conveniences) as a concession to implied town planning that taste demanded.

2.7 NATIONAL AND LOCAL TAXATION

From the moment men decided to live in a community, however small, it was inevitable that each had to make some contribution for the common good of all and inevitably at some sacrifice to the individual's private source of goods or energy. As communities became larger and more 'civilized', contributions towards the common good were necessarily collected in currency rather than in any other form and the mixture of that which was voluntary and that which was imposed merged into mandatory taxation. In times of threats to regional or national security, money for defence and ancillary matters can be raised by imposing taxes on the members of the community, often in the form of a percentage of the produce of the land whether this be crops or livestock, but in particular a levy can often be made based on some portion of the value of the land itself, including the buildings on it.

Dating from the Norman conquest, land tenure originated from the acquisition of land by conquest and its fragmentation by the head of the conquering army to his followers, to be held by them with no further charge other than knight service. The land so given had to be cultivated and with the gradual evolution from serfdom to tenantry, taxes on the tenantry were an obvious means open to the State to extract revenue which could not be obtained from the Lord of the Manor, whose grant was initially free from monetary payments.

The four maxims for taxation in general laid down by Adam Smith[36] still remain valid but doubtless had he written in the second half of the twentieth century, he might well have added a fifth! His four maxims can be summarized as:

1. Each individual should be taxed in proportion to his revenue.
2. The tax should be certain and not arbitrary.
3. The tax should be levied at a time and in a manner convenient for the contributor to pay it.
4. Every tax should be contrived to take out and keep out of the pockets of the people as little as possible over and above what it brings in to the public treasury.

Adam Smith explains that the 'levying of it may require a great number of officers, whose salaries may eat up the greater part of the produce of the tax, and whose perquisites may impose another additional tax. It may obstruct the industry of the people and discourage them from applying to certain branches of business which might give maintenance and employment to great multitudes. By the forfeitures and other penalties which those unfortunate individuals incur who attempt unsuccessfully to evade the tax it may frequently ruin them.' He goes on to suggest that 'an injudicious tax offers a great temptation to smuggling,' and that 'the law, contrary to all the ordinary principles of justice, first creates the temptation and then punishes those who yield to it,' and adds that 'by subjecting the people to the frequent visits and the odious examinations of the tax-gatherers, it may expose them to much unnecessary trouble, vexation and oppression, and though vexation is not, strictly speaking, expense, it is certainly equivalent to the expense at which every man would be willing to redeem himself from it.'

There seems little evidence in his treatise that Adam Smith considered the concept of taxation for socio-political purposes, but those taxes of the type he knew, and others of which he had no prescience, have undoubtedly affected property transactions and development. In this connection, it is reasonable to assume that Governments before the twentieth century looked upon taxation for the raising of revenue to defray the expense of actual war (or the threat of it) and for the administration of internal affairs, more or less following the four maxims postulated by Adam Smith. The fifth maxim he might have laid down will be dealt with in the ensuing commentary on income tax, Capital Transfer Tax and others, which appear to have other objectives.

Various taxes have affected property owners in the long history from the Norman conquest to the present day and whilst mention must be made in passing of some of those that had minimal effect, two of the more fanciful taxes that have been imposed on property in the past include a hearth tax and a window tax.

(a) Hearth tax

This was a tax at a rate of 2 shillings for each hearth; it was levied on all houses unless the occupier (a) was exempt from paying church or poor rates or (b) was certified as living in a tenement under the value of 20 shillings per annum, and not having land to that value, nor possessing goods to the value of £10. It was first levied in 1662, but owing to its unpopularity, was repealed in 1689, although producing £170 000 a year. The principle of the tax was not new in the history of taxation, for in Anglo-Saxon times, the King derived a part of his revenue from a 'fumage' or tax on smoke, levied on all hearths except those of the very poor.

(b) Window tax

This was a tax first levied in England in 1696, for the purpose of making up the deficiency arising from clipped and defaced coin in the recoinage of silver during the reign of William III. All houses inhabited except those not paying Church or poor rates were assessed 2 shillings a year. An added tax was levied according to the number of windows: on 10 -19 windows the additional tax was 4 shillings. In its first year the tax raised £1.2 million. It was increased six times between 1747 and 1808 but was reduced in 1823. After strong protests in the winter of 1850—51, it was repealed on 24 July, 1851, and replaced by a tax on inhabited houses.

Visible evidence of the effects of the window tax is apparent in many streets in London where older houses still remain, especially in those that stand on corner sites and, therefore, have return frontages. The upper windows of these return frontages have often had window frames removed and the apertures filled in with bricks, sometimes plastered over. After the repeal of the tax these apertures were not restored to their former state, either because the owners had grown accustomed to being without the extra light or because the interior of the rooms affected had been redesigned and reopening the apertures would have been unsightly. It is interesting to have at least two examples of the amounts collected in tax and from that most informative of sources, Thomas Paine[37] we learn that the returns for 1778 showed:

Houses and windows by the Act of 1766	£385459 11s 7d
Houses and windows by the Act of 1779	130739 14s 5½d
Total	£516199 6s 0½d

According to the same source, the amount of land tax for the year ending Michaelmas 1778 was £1950000 out of the total tax revenue from all sources of £15572970. Customs and Excise Tax was over £10 million for that year. However, the main examples of tax imposed prior to the 1914–18 War are dealt with under the headings of:

Land tax
Income tax
Death duties
Local taxation (rates)

2.7.1 Land tax

This tax dates from 1692 and was developed from the monthly assessments under which Parliament specified for each county the total sum required and each county raised its contribution by a rate based on the yearly value of all property, real and personal, and on the income from certain 'offices.' In intention, the land tax was a combination of property (real and personal) and income tax, but personalty gradually escaped assessment. In 1798, William Pitt made the quota then due from each area a perpetual charge on the landed property of that area, but permitting a property owner to redeem the liability on meeting a capital payment, thus the land tax became converted into a redeemable rent charge on landed property. Each parish had its quota to be raised annually and this was levied by a rate not exceeding

1 shilling in the £1 on the annual value of all land and buildings in the parish. (Subsequently an owner of the land could redeem the charge by payment of a capital sum equal to twenty-five times the annual tax assessed for the year ended 24 March 1940). Where redemption was effected the land was exonerated from the tax and the quota of the parish was correspondingly reduced. Because the annual value of any buildings erected on the land came into the annual value for the charge to land tax, redemption generally took place before building operations started.

Changes in the value of money made what was a small annual collection progressively unimportant to the revenue. The effect of the tax on development was likewise very small indeed, since even at twenty-five years' purchase, the redemption sum made little impact on the financial viability of a development and there is no evidence that it prevented development or redevelopment unlike the development tax imposed in recent times.

It may be of interest to record that recent legislation on property taxation was to some extent anticipated as long ago as 1791 by Thomas Paine who attacked the redemption tax in his book, *The Rights of Man*[37]. He offered a 'plan for its abolition by substituting another in its place,' pointing out that the amount of the tax collected in 1788 was a mere £771657.

> 'When taxes are proposed the country is amused by the plausible language of taxing business. Admitting — a thousand pounds is necessary for the support of a family, consequently, the second thousand is of the nature of a luxury and so by proceeding — arrive at a sum that may not improperly be called a prohibitable luxury.'

He admits it 'would be impolite to set bounds to property acquired by industry — but there ought to be a limit to property — by bequest.' His book contains a table modified here as Table 2.3, which, he proposed, would 'supersede the aristocratical law of primogeniture' if it became the basis of taxation. Sir Stafford Cripps, the Chancellor of the Exchequer in the immediate post-Second World War Government imposed a levy on high incomes that took the tax above 20 shillings in the £1.

Table 2.3 A tax on all estates of the clear yearly value of £50 and up, after deducting the land tax

	s d per pound
To £500	0 3
From £500 to £1000	0 6
On the 2nd £1000	0 9
On the 3rd £1000	1 0
On the 4th £1000	1 6
On the 5th £1000	2 0
On the 6th £1000	3 0
On the 7th £1000	4 0
On the 8th £1000	5 0
On the 9th £1000	6 0
On the 10th £1000	7 0
On the 11th £1000	8 0
On the 12th £1000	9 0
On the 13th £1000	10 0
On the 14th £1000	11 0
On the 15th £1000	12 0
On the 16th £1000	13 0
On the 17th £1000	14 0
On the 18th £1000	15 0
On the 19th £1000	16 0
On the 20th £1000	17 0
On the 21st £1000	18 0
On the 22nd £1000	19 0
On the 23rd £1000	20 0

2.7.2 Income tax

(a) Introduction

Any goods, wares and merchandise or other chattels of the personal estate 'whatsoever within this realm' were taxed at four shillings in the £1 according to their true yearly value by the Land Tax Act 1692. In addition the Act imposed a similar tax upon profits and salaries of any office or employment by profit, excepting only members of the armed forces. The amount of duty on personal estate or its proportion in the total land tax raised is unfortunately not on record, but what is known is that over the course of years, the yield on the tax dwindled almost to nothing.

Income tax as such was first introduced in 1799 by William Pitt as a 'temporary' tax to provide finance for the war with France. The tax payer was asked to state his income from all sources but incomes under £60 were exempt. The general rate of tax was 10% on all incomes of £200 upwards. For incomes between £60—£200 the rate was graduated with allowances for children and repairs to property amongst other items. Students of the Napoleonic Wars will recollect that the Treaty of Amiens in 1802 temporarily brought in an era of peace and in consequence the 1799 Act was repealed in May of that year. When war broke out again in 1803 a new income tax Act was passed which was so soundly based as to remain in essence the framework of income tax law to the present time.

The sources of income were divided into five sections, which came to be known officially as Schedules A, B, C, D and E. The principle of collection of the taxes at source was introduced. Former statements of income were abandoned and in their place were substituted a number of required statements of income from particular sources. This new method of collection ensured that despite the reduction of the 1799 rate of 10% to 5%, the amount collected was almost equal to that of the earlier Act. However, there were further Acts of 1805 and 1806 which modified the 1803 Act and the 1806 Act remained in force until 1815, for some time after the Battle of Waterloo. It then automatically expired in 1815 *and from 1816 to 1842 there was no income tax payable whatsoever.*

Following a period of economic recession in 1842 Sir Robert Peel re-introduced the tax in order to reduce the import duties on a number of articles 'which entered into manufactures as chief constituent materials.' As with the 1806 Act, it was intended to be a temporary measure for a period of three years only. Sadly the Act was never repealed and subsequent Acts can be said to be mere modifications of it. The 1842 Act raised the exemption limit to £150 per annum — the 1853 Act brought it back to £100. The Act of 1876 restored it once again to £150, the 1894 Act to £160 and it remained at this figure until the outbreak of war in 1914, followed by the Act of 1915 which altered the figure to £130. For purposes of comparison it should be noted that at the date of its demise in 1815, the tax was 2 shillings in the £1; from 1842 to 1900 it varied between 6d and 8d, although it was as low as 2d, in the years 1874 to 1875. The South African war caused the rate to go just above 1 shilling and during the 1914–18 War it touched 6 shillings.

(b) **The Schedules**

The Act of 1799 which required each tax payer to make a return of his total income was thought to be a gross infringement of the liberty of the subject. In 1803 the Chancellor of the Exchequer, Addington, was responsible for the introduction of the system of classifying income by five types of Schedules. To overcome prejudice he organized separate returns of income under each Schedule to be sent to a different officer so that no one officer should know the total of any man's income.

Of the current Schedules, F imposes a charge to tax on distribution and was introduced in 1966. Schedule A, as originally enacted, was not only a tax on income from rents actually received but was also levied on owner-occupiers on the notional income that could be obtained from the property if it were let to an occupier who was not the owner. (In this form it was abolished in 1963). Schedule B deals with profits arising from the occupation of commercial woodlands. Schedule C deals with interest on public funds, Schedule D on overseas securities other than public debts and trades or professions and Schedule E deals with income from employment.

2.7.3 Supertax and surtax

In addition to income tax a new tax was introduced in 1909 called supertax which was levied at the rate of 6d in the pound on incomes exceeding £3000. By 1914, the top rate of this supertax was 1s 3d. By the end of the 1914–18 war, this rate had increased to 4s 6d, and in percentage terms it meant that a tax structure which in 1914 maintained a maximum rate of income plus supertax at 12½%, escalated by the end of the war to 52½%. The term 'supertax' was later superseded by the term 'surtax'. The technical effect of the change in terminology was to make surtax a part of income tax but payable a year later.

2.7.4 Death duties

The history of death duties commences with the Stamp Act of 1694 and closes with the Finance Act of March 1975. The changes in the way in which duty of one sort or another was charged can best be illustrated by a summary of the legislation.

(a) 1694 — Stamp duty
A flat rate duty payable to the State before a will could be proved or letters of administration granted. The Act was described as a 'probate and administration duty.'

(b) 1780 — Legacy duty
This was an extension of the 1694 Stamp Act and was in the form of a Stamp to be attached to receipts for any bequest of an intestate succession to personalty.

(c) 1853 — Succession duty
This was a duty on all types of property other than property already charged to legacy duty.

(d) 1881 — Account duty
A duty (strangely portending the capital transfer tax of nearly 100 years later) whereby gifts made less than three months before death came within the tax net.

(e) 1885 — Corporation duty
A tax on the annual income of a corporation which because it was deemed immortal accordingly could never be assessed for death duty. The yield from this tax proved negligible in view of the many exemptions. A curious and now defunct system operated to share the duty between central and local Government and the Bank of England created special local taxation accounts for this purpose.

(f) 1889 — Goschen's 'estate duty'
Now known as the 'temporary estate duty,' to distinguish it from the 1894 version that followed, was a 1% flat rate charge on all estates over £10000 in value.

(g) 1894 — Finance act
Estate duty was imposed by the Finance Act 1894 in sub-

stantially the form which existed until 1974. Duty was payable by all persons dying after 1 August, 1894. It superseded probate duty and account duty and made chargeable all property passing on death, without regard to the relationship of a beneficiary to the deceased owner. There was a provision that duty was payable on *all* property passing on the death of the deceased and brought into taxation not only absolute ownerships but included real estate not formerly bearing probate duties and also limited interests and even property in which there was no interest at all but deemed to pass, i.e. settled estates. It was the first time that the principle of abrogation was attacked.

(h) **Lloyd George's Budget of 1909**

By any standards the Finance Act of 1909 must be considered a frontal attack on the bastions of the land and property owners. It introduced four duties on land and, understandably, was bitterly opposed by those affected.

Increment value duty

This was levied on every 'occasion' of sale or transfer of land at the rate of 20% on the increase in the site value of the land accruing after 30 April 1909. Provision was made for periodical assessments every fifteenth year in the case of land held by corporate bodies.

Reversion duty

This was levied at the rate of 10% on the value of the benefit accruing to the lessor on the termination of long leases, the benefit being in general the difference between the value at the beginning and at the end of the lease.

Undeveloped land duty

This was an annual duty of ½d in the £1 on the site value of undeveloped land, i.e. land not being used for agriculture, business or building purposes, etc., and was intended to fall on land ripe for building, gardens, public parks, recreation grounds and woodlands being exempt.

Mineral rights duty

This was an annual duty of 1s in the £1 on mineral royalties, wayleaves, etc. In order to levy the tax it was necessary to value all lands and buildings, as at 30 April 1909 — a colossal task involving the valuation of 11 million units of land including, in particular, the determination of 'site value' which represented, broadly speaking, the market value of the land when divested of buildings, trees or other improvements upon it. The work of valuation proceeded for five years and was practically completed by the outbreak of World War I, the total being estimated at £5250 million.

The first three duties yielded a small though increasing revenue in the early years. Serious difficulties, however, soon arose in connection with the assessment of the land values required by the Statute and these problems ultimately led to the practical suspension of the undeveloped land duty and reversion duty and also the increment value duty. The Revenue Bill of 1914 was intended to remedy the situation but the outbreak of war prevented the passage of the Bill. At the end of the war it was decided to repeal these three duties. This was accomplished by the Finance Act of 1920, and provision was made to repay all the duty collected after their imposition, only the mineral rights duty survived.

So far as the effect of this attempt to impose a tax on property development was concerned, it is obvious that the administration set up to collect the tax broke down, and whereas it is difficult to the point of impossibility to collect evidence to support the contention, it would be reasonable in view of the admitted figures of tax collected, to assume that it had a minimal effect on development or redevelopment.

2.7.5 Rates — the local taxation system

In London and the principal provincial cities, substantial office blocks and the larger factory complexes command rentals of several hundred thousands of pounds a year. The rates payable often equal the rent and, to give an indication of the situation, it is possible to calculate that the headquarters of a City of London- based organization occupying 100 000 square feet at an annual rent of £10 per square foot could be paying close on £1 million per annum in rates plus another 8% for water rates. This latter figure amounting to £80 000 would seem an extravagant price to pay for the ablutions and tea-making facilities provided for the work-force.

The history of rating reveals that its origins were national rather than local. They arose out of the feudal system imposed by William I who, after the Norman Conquest, gave lands to his knights. In return he required attendance at the Hundreds County Courts and Parliaments. As mentioned earlier, knight service was a form of land tenure and the holder was liable to provide an armed and mounted knight for military service for every £20 worth of land held. This could be varied to locage tenure, i.e. paying £1 per annum for every £20 of land, the proceeds going to the Exchequer to defray the expenses of Government administration. As time passed the amounts collected fell short of national requirements and it was decided by Henry II to devise a system whereby each locality elected a jury whose duty was to assess real and personal property values on which an annual tax was levied amounting to one-fifteenth for county, and one-tenth for borough inhabitants. These taxes did not cover expenditure of a purely local nature, such as the building and maintenance of roads, bridges, sea walls, etc., for which the Lord of the Manor and the more affluent residents took responsibility, either providing labour and materials or money.

What started as a purely voluntary system was eventually superseded by local Government officials and the contributions previously made voluntarily were then exacted as of right arising out of custom and their collection could be legally enforceable. In particular, contributions could be enforceable against an inhabitant who benefited by reason of his land adjoining his neighbours where, for instance, they had constructed a wall to prevent flooding or where a bridge had been built to give better access for all.

Custom needed to be reinforced by legislation as both population and economic growth necessitated an expansion of road networks, bridges, sewage and the like; and it was the last of these items that provided the first rating legislation. Entitled the Statute of Sewers it was enacted in 1427 and remained on the Statute Book until its successor The Bill of Sewers was enacted in 1531.

From these beginnings Acts were passed dealing with local issues in their greatest numbers between 1700 and 1800 (an era noted for the quantity of private Acts passed by Parliament). The Sewage Acts of 1427 and 1531, although of incidental historic interest, are of minor importance to the pattern of legislation that intensified with the passage of time. This pattern can be more easily followed under the headings of

The relief of destitution
The county rate
The borough rate
The highway rate
The public health rate (including the police rate)
The education rate
The London rate

Arising out of the legislation were questions relating to rateability, that is to say the liability to pay rates, and the basis of their assessment. These questions involved both Parliament and the Courts before they were answered and, in particular, what had to be settled was whether both the liability to pay and the assessment should be related to the value of the hereditament or to the financial ability of the ratepayer.

(a) The relief of destitution

This can be summarized by the following legislation:

1388	To control beggars!
1427	Statute of Sewers -- to provide sewers.
1531	Replacement of 1427 Statute.
1536 and 1539	Relief of destitution.
1572	Authority to administer 1536 and 1539 provisions by Justices of the Peace.
1597	Act to consolidate the above.
1601	The Poor Relief Act gave authority to Justices of the Peace to institute a uniform basis of liability and fix the frequency of

	making rates. Furthermore it established the rate as the basis for other payments.
1834	The Poor Law Amendment Act established work houses and controlled outdoor relief, appointed three Commissioners, amalgamated parishes into Unions, provided for election of Guardians of the poor. Each parish was made responsible for its own poor.
1862	New provisions enacted to control Guardians of the poor.
1865	Responsibility for parishes to provide outdoor relief abolished.

(b) **The county rate**

This was instituted to levy a rate on each parish as a whole and not on individual ratepayers.

1739	A Consolidation Act to bring in 1601 type legislation to incorporate hitherto separate rates for roads, bridges, prisons, etc., into one single rate.
1815	An amending Act — The County Rates Act — authorizing Justices of the Peace to make a county rate assessing parishes on a poundage basis.
1888	An Act transferring the authority of making a county rate from Justices of the Peace to the county authorities. (A system continued till the abolition of the county rate in 1948).

(c) **The borough rate**

1835	The Municipal Reform Act confirmed the poor law assessments as the basis for borough rates and provided for the poor rate to be collected with the borough rate. (By this Act the borough became a Sanitary Authority.)

(d) **The highway rate**

1555 and 1662	A Highways Act to repair roads, following previous Acts naming particular roads in specific areas.
1691	An Act authorizing general highway rates.
1767	A Consolidating Act.
1835	A further Consolidating Act giving powers to the Surveyor of highways.
1848	An Act for the creation of Boards of Health which replaced the functions of the Surveyor of highways.

(e) **The public health rate (including the police rate)**
The provisional maintenance of paving and street cleaning, sewers, street lighting and watch committees was the subject of independent action by local authorities, and private Bills were brought before Parliament on numerous occasions during the eighteenth century.

1830	The Lighting and Watching Act provided a measure of uniformity.
1833	The repeal and re-enactment of the 1830 Act providing for inspectors and the inclusion of the rate with the poor rate.
1848	The Public Health Act was the first Act dealing with sanitation, water and other connected items and running with the poor rate.
1858 and 1875	A Public Health Act of 1858 was amended by the 1875 Act dividing England and Wales into Urban or Rural District Councils including all former authorities. This reduced rating to (a) a general district rate; (b) a highway rate and (c) a private improvement rate for urban districts. Rural district rating was similarly reduced to (a) a general district rate and (b) a special expenses rate, both collected with the poor rate.
1829	The Metropolitan Police Act allowed the rating authority to impose a maximum of 8d in the £1 for police purposes.

1835	The Initial Corporations Act set up Borough and County police forces.
1839	The County Police Act made the formation of a force compulsory for counties.
1856	County and Borough Police Act extended powers to boroughs.
1888	Local Government Act — made compulsory setting up of joint standing committees of Justices of the Peace and County Councils. Provincial police costs were in addition to that included as part of the borough or county rate and the poor rate.

(f) The education rate

1870	The Elementary Education Act provided for school boards' expenses when voluntary schools were insufficient.
1889	Technical Instruction Act provided for technical and trades instruction outside the scope of elementary schools. The maximum rate was fixed at a penny in the £1.
1902	Education Act abolished school boards substituting county or country borough administration at the three levels of elementary, secondary and technical education. County Borough Councils with over 10000 population and urban district councils with over 20000 could take education costs directly out of their respective general rates.

(g) The London rate

1899	The General Rate Act — the first step to consolidate separate rates into one single rate. The Act provided for metropolitan borough councils to take over the functions of the Overseers of the poor.
1901	The London Scheme amended the General Rate Act, incidentally excluding the City

which collected a general rate and a poor rate separately.

(h) Rateability and the basis of assessment

From the inception of the system the liability for rates posed questions which at first the Courts answered by assessing the ratepayer according to his ability to pay.

Debates about what was rateable engaged the Courts continually and a number of decisions were handed down which determined that stock-in-trade and other goods were caught in the system. The more significant of the Acts and Court decisions are listed below.

1572	14 Elizabeth I, c.5 — a general tax to provide alms for the poor imposed on inhabitants.
1589	The Jeffreys Case — Judge Jeffreys decided that although a resident of Chiddingley, the ratepayer was liable to contribute to repairing Hailsham Church since he owned land in that parish.
1597	An Act making both inhabitants and occupiers liable for rates for mines, woodlands and sporting rights, plant and machinery.
1633	A case heard by Sir Robert Heath C.J. relying on Statute 43 Elizabeth I, c.2. His judgement reads that the ratepayers 'shall be by ability or occupation of land or both and whether the visible ability of the parish where he lives or general ability whereso-ever and whether his rent received within the parish where he lives shall be accounted visible ability, and whether he shall be taxed for them only and for any rent received from other parishioners and what shall be said visible about it land within each parish is to be taxed at the charges in the first place equally and indifferently but there may be an addition to the *personal visible ability* of the parishioners within that parish according to good description.'

1633 Sir Anthony Earby's Case determined that 'assessments ought to be made according to the visible estate of the habitants there (in Boston) both personal and real and that an inhabitant there is to be taxed by them to contribute to the relief of the poor in regard to any estate he had elsewhere in any other town or place but only in regard of the visible estate he hath in the town where he doth dwell, and not for any other land which he hath in any other place or town.'

1635 In Dalton's *Country Justice*, 1635 edition (reprinted 1727), page 218, it was stated 'in these taxations there must be consideration. . . to equality and then to estate. Equality that men may be equally rated with their neighbours and according to an equal proportion. Estates that men be rated according to their estates, goods known or according to their known yearly value of their lands, farms or occupyings but not by estimation supposition or report. Also herein the charge of family retinue etc., is in some measure to be regarded for if one valued at £500 in goods hath but himself and his wife and another estimated £1,000 hath wife and many children etc., the first man by reason is to be rated as much as the other and so of lands.'

1684 According to Hale C. J. in his posthumous tracts published in 1684 'because those places where there are most poor, consist for the most part of tradesmen whose estates lie principally in their stocks, which they will not endure to be searched into to make them contribute to raise any considerable stock for the poor, nor indeed so much as to the ordinary contribution but they lay all the rates to the poor upon the rents of lands and houses which alone without the help of stocks are not able to raise the stock for the poor.'

Although there was obvious sympathy for the concept of ability to pay nevertheless the Courts found it easier to apply that of the annual value as the basis of assessment.

1698 A case heard in the Court of the King's Bench (Comb 478) produced in the judgement the statement that 'the rent is no standing rule for circumstances differ and there ought to be regard *ad statum et facultatus*', i.e. personal ability to pay.

1706 *Rex* v. *Barking (Inhabitants)*. It was held that the stock-in-trade of a farmer was not rateable.

1777 *Rex* v. *Andover Church Wardens*. The interpretation of 'inhabitant', the word used in the Poor Relief Act of 1601, was held to include houses, saleable growing timber and mines. The basis of assessment was the annual rental value. This coincides with the 1633 Sir Anthony Earby case, the 1634 Dalton's *Country Justice* case and the 'posthumous discourse' of Justice Hale.

1798 *Rex* v. *Skingle* 7 TR 549. Here the annual value was held to be the proper basis for assessment and not the actual rental paid.

1827 *Rex* v. *Attwood* (1827), 6B and C 227. A mine owner in occupation was assessed at the net rental or royalty receipts calculated after allowing for outgoings to maintain the mine but not to take account of capital expenditure applicable to all types of rateable property.

1829 *Rex* v. *The Duke of Bridgewater's Trustees* (1829), 9b and C 68. A canal was declared to be properly assessed at the rent at which it would let and not the gross receipts less expenses. However, in computing the rent, allowance was made for such repairs as to make the occupation of the hereditament productive This judgement presaged the definition of the 'hypothetical tenancy' which, with no significant alteration by the Acts of 1836, 1862 and 1925, persists to the present time.

1829 The Advertising Station Rating Act enacted to rate the owner of a house who let a flank wall for advertising separately from the house, the tenant of which would not be charged.

1831 *Rex* v. *Chaplin* (1831), 1b and Ad 926. This concerned a canal which was assessed on the sum of interest payable under a mortgage, this being regarded as equivalent to the rental value. The judgement is of some note since it introduced the concept of determining value on the basis of capital value or the interest thereon.

1832 In 1832 it was decided that land must be rated in proportion to the rent which the tenant at rack rent would undertake to discharge all rates charges and outgoings including the sewer rate (*Rex* v. *Adams* (1832), 4b Ad 61.

1835 *Rex* v. *Woking (Inhabitants).* This case hung on those items that could be deducted from gross receipts to arrive at the net annual value. Thus compensation paid to persons injured as a result of the operation of a canal was not allowed.

1836 The Parochial Assessments Acts (see the Union Committee Act 1862).

1839 *Rex* v. *Lunsdaine.* In spite of legislation to the contrary (The Parochial Assessments Act 1836) the Judge held that stock-in-trade was rateable. Subject to the 'ability to pay' the general practice was to fix the sum at 10% of the rate applied to land. No record is available of any appeal against the decision.

1840 The Poor Rate Exemption Act. This exempted machinery and stock-in-trade from assessment. It was enacted to stop the Poor Law Commissioners, through their overseers, from rating stock-in-trade and profits. It contained the provision that it 'shall in no ways affect the liability of any parson or vicar or of any occupier of lands, houses, tithes, impropriations of tithes, coalmines or saleable underwoods to be taxed for and towards the relief of the poor.'

1862 The Union Assessment Committee Act of 1862 defined gross estimated rental as 'the rent at which the hereditament might reasonably expect to let from year to year free of all usual tenants rates and taxes and tithes commutation, and rent charge, if any, provided nothing herein contained shall repeal or interfere with the provision contained in Section I of the Parochial Assessments Acts 1836 defining the net annual value of the hereditaments to be rated.' The object of the Act was to reduce argument on the difference between 'gross estimated rental' and net annual value. The basis of assessment was that since most lettings were effected with landlords undertaking repair the gross estimated rental corresponded to actual transactions. It still left unanswered the question of how to assess hereditaments such as mines and owner-occupied factories, public undertakings and others where the landlord was not responsible for repairs. In such cases the profit basis was used to obtain a rateable value (net) and then add a percentage for repairs to get to a gross value. There was no scale of deductions to obtain a rateable value from a gross figure thus there were variations in different geographical areas.

1869 The Valuation Metropolis Act. This Act applied to London alone. Its main significance was in the redefining of terms used in the 1836 Act altering 'gross *estimated* value' to 'gross *annual* value' and defining it as 'the annual rent which a tenant might reasonably be expected taking one year with another to pay for an hereditament if the tenant undertook to pay all usual tenants rates and taxes and tithes commutation and rent charge, if any, and if the landlord undertook to pay the cost of repairs and insurance and the other expenses, if any, necessary to maintain the hereditament in the state to command that rent.'

1874 A House of Lords ruling that mines were rateable.

2.8. POWERED TRANSPORT AND ITS EFFECT ON DEVELOPMENT

The nineteenth century American writer O. Henry tells a delightfully romantic story of a young man anxious to prevent his sweetheart leaving New York by a trans-Atlantic liner. Henry arranges for him to hire several dozen cabs to converge so as to hem in the carriage taking the young lady to the docks. The subsequent traffic jam gives the young man time enough to make a proposal that upsets her previously arranged travel programme, not to mention the inconvenience to all other travellers in the city — a point on which Henry does not dwell.

There was nothing romantic about the Royal Commission that was set up in 1905 to enquire into the serious economic cost on business in general that arose from traffic delays. The aggregate of stoppages per day during the twelve-hour period from 8.00 a.m. to 8.00 p.m. shows that something like 15500 vehicles proceeded down St James's Street during that time, going from east to west experiencing 230 delays, and the total of these delays was no less than one hour and 47.3 minutes. At the Bank of England junction, where Princes Street, King William Street and Queen Victoria Street meet, there were up to 337 delays occasioned by some 3000 vehicles passing. The Commission found that in the interests of public health, in addition to making decent housing possible, it was imperative to improve transport facilities. Additionally, it gave as one of the considerations for improvement 'the prompt transaction of business'. The report suggested that comprehensive plans for the improvements of streets and main roads be continuously improved as financial considerations allowed, since it gave the narrowness of the streets as the main reason for the delays.

In 1905, the number of cars on the roads of Britain totalled 15895. In addition, there were 7491 vehicles described as hackneys and 9000 goods vehicles. The total number of mechanically-propelled road vehicles was thus under 33 000. By 1925 this figure was near enough 1.5 million and had climbed to over 15 million by the 1970s. The development of, and trends in, urban and suburban transport, and the effect on the pattern of property development is best illustrated by examining the effect on London and its environs, although similar events affected other major cities.

2.8.1 The major influences of changes in motive power

The importance of transport on the development of a country as a whole, and not just related to property development, is shown by the location, from earliest times, of the principal towns and cities devoted to receiving ships from abroad. The nature of the geographical advantages of some locations is undoubtedly responsible for the growth of particular cities. It would not be unreasonable to assume that water-borne transport played the most dominant role in the development of these islands until well into the twentieth century.

Inland, preceding the railways, the canal system seemed a natural adjunct to the existing method of transporting coal and iron, amongst other things, from one seaport town to another by coastal vessels. By the middle of the nineteenth century, the advent of the railway system made it possible to transport goods much more quickly as steam replaced sail on the sea and the horse on the tow-path. Coincidental with the effects of the Industrial Revolution, the railway undoubtedly transformed inland towns and cities and their growth as industrial centres inflated the working population, for whom surburban housing was soon required in great quantity. This led to a demand for transport from home to work and although some towns and cities had electric tramways, for the most part local transport was horse-drawn. The motor car or motor omnibus did not appear on the scene until just before the First World War and mechanical transport was confined to railway systems.

2.8.2 Roads, canals and railways

In considering the relationship between the roads, canals and railways, it must be remembered that before the eighteenth century, it would not have been possible either technically or economically to construct a good road system in Great Britain. There was a complete lack of technical skill in road construction and although this might have been acquired by experience, as occurred at the time of the Roman occupation, there were economic forces to be reckoned with which did not exist when the Roman road network was developed. As the Romans advanced, they drew their labour from the conquered Britons. The land and materials were seized by force and without compensation to their owners and, furthermore, the Romans had a military necessity.

Before the eighteenth century, there was not sufficient wealth or population to warrant the establishment of a network of good roads and such roads as were in existence connecting important towns were often impassable on account of mud in wet weather. The eighteenth century road makers expected that water would flow to the sides from a steep slope, but in creating such a slope they made it necessary for vehicles to keep to the crown of the road. Consequently, vehicles travelling in all directions used only the crown, deep ruts were caused very soon, which generated further mud. The result was that goods were conveyed on pack horses more frequently than by horse wagon and obviously, therefore, those who required heavy loads to be carried for long distances looked to the rivers or coastal vessels plying from one port to another.

(a) Turnpikes and tolls

Characteristically, the eighteenth century with its emphasis on private enterprise, solved the difficulty of maintenance of roads by granting to Turnpike Trusts the right to control stretches of road, on condition that they put them in order and maintained them satisfactorily. Every person, animal and vehicle using the roads had to pay a toll and the money so received was used for keeping the roads in repair and paying the toll-gate keepers' wages. This set up the principle that the cost of the maintenance of the roads should be a charge upon their users and not upon the public at large.

The first Turnpike Act was passed in 1663, and the system became increasingly common as time went by with over 450 Turnpike Acts being passed between 1760 and 1774. At the close of the eighteenth century, and in the early part of the nineteenth century, John Metcalf, Thomas Telford and John MacAdam, three engineers whose names are remembered to this day, improved the quality of the surfacing of these roads. During the first half of the nineteenth century, the Turnpike Trusts amalgamated and roadways generally became very much improved. The advent of the steam railway, however, diminished the revenue of the Trusts and as their revenue disappeared and they were unable to maintain the roads properly, local authorities took over the duty of repairing the roads.

(b) Canals

The improvement to the road system, however, did not come in time to deal with the Industrial Revolution, and the need to convey goods was met by the construction of canals. One of the earliest of these, completed before the end of the seventeenth century, connected Aire and Calder. In 1720, a straight course was cut across a number of loops in the rivers Irwell and Mersey, which had the effect of opening up waterways between Liverpool and Manchester. A significant venture was the Bridgewater Canal of 1759, connecting Manchester with the colliery of Worsley. This was intended to facilitate the carriage of coal from the mine to the city.

It was in every way successful and had the effect of providing cheap coal for Manchester. Between 1760 and 1797, an enormous number of canals was cut. As they were all constructed by private enterprise with no central planning, the canals varied in width, depth, overhead clearance and dimensions of locks, making it impossible for barges over twenty tons to traverse the whole system. The canals imitated the Turnpike Trusts in that they did not undertake the carriage of goods, but merely provided a route to anybody who would pay the required toll fees.

Despite their obvious defects, the canals did provide a better system of transport than any that had been available before they were constructed, and without them industrial development would have been impossible. Food was conveyed by them from the south and east to the north and west. Clay from Cornwall was transferred by canal to the potteries of the country and to the ports, and the problem of moving coal was solved entirely in this way. A large amount of traffic which had previously been brought from port to port by coast-hugging vehicles was now sent by canal. As with the toll roads, the fortunes of the canal companies dwindled with the coming of the railways. Railway trucks could be run direct from the pithead to the wharf, and the work of bunkering ships was simplified to an extent which was impossible when coal was conveyed by canal.

Many of the canal companies were, in fact, purchased by the railway companies, who were forced to buy them out when the former opposed new railway routes. In 1888, acquisition of canals by railway companies without statutory authority was prohibited. Undoubtedly, the Government of the day viewed with some concern the effects of railway acquisitions of canals. After purchasing them to quash opposion to new railway routes the canals were left to fall into disuse and it was considered at the time that some trade and possibly a hitherto useful means of a public transport facility might have been lost unnecessarily.

One notable exception to the general trend was the opening of the Manchester Ship Canal, which took six years to build. Some 36 miles in length, it allows ocean-going ships to come up to Manchester with their cargoes. Such is the success of this canal that Manchester still ranks as the third largest port in the United Kingdom. Such canals that have remained in operation passed under the control of the Docks and Inland Waterways Executive of the Transport Commission under the Transport Act of 1947. An Act of 1953 renamed the authority, and a Transport Act of 1962 assigned the control to the British Waterways Board.

(c) Railways

The earliest railways were short private lines connecting collieries with a river or canal. As early as 1630, efforts to improve the rutted colliery road where the wagons were hauled by horses or men, resulted in flat or grooved wooden plates being imbedded in the roads. Iron plates were used in place of wood after 1738 and in 1767 cast-iron rails were used.

William Jessop invented the idea of using trucks with flanged wheels and these were used between Wandsworth and Croydon by the Surrey Iron Railway which, under an Act of Parliament passed in 1801, operated for the conveyance of coal and corn, etc. The trucks were hauled by horses, but experiments were conducted in the use of cables which could be wound around drums attached to a stationary steam engine. The early steam engine or steam coaches, as they were called, produced by George Stephenson and other inventors, were not railway engines but ran on the public roads.

The first railway, as it came to be known, was the Stockton and Darlington Railway of 1821. Although the line was operating by 1825 with an engine to Stephenson's design, the fact remains that it was the original intention of the promoters to use horses for haulage. The history of the development of the railway system in England proceeded apace, almost every major town being connected to the system, with London particularly well served. By 1850, England was in possession of a vast network of railway lines, and as construction progressed there was a great tendency towards consolidation; the gauge was standardized and, more importantly, the Government began to take a greater interest in them. An Act of 1840 gave powers to the Board of Trade to inspect and supervise the opening of any new line and to order its closure if it was dissatisfied with its condition.

Railway dividends rose to 10%, profits in excess of this

had to be devoted to the reduction of rates and fares. Railway Company byelaws had to be submitted for approval and although it did not do so under the 1840 Act, the Government did take powers under certain conditions, and after a set period of time, to purchase the lines. The Cheap Trains Act of 1844 directed all railway companies to run one train daily in each direction, calling at all stations in which passengers could be conveyed, at fares not exceeding 1d per mile. A year later, Parliament fixed maximum rates for the carriage of goods. The Railway and Canal Act 1854, commonly known as 'Cardwell's Act' directed railway companies to afford reasonable facilities for the carriage of goods and forbade the giving of preferences. The Act thus removed some corruption whereby railway companies could quote differing rates to rival manufacturers, ruining some and making fortunes for others.

Public control was extended with the passing of the Railway and Canal Traffic Act 1888, which was mainly concerned with the question of rates charged for the carriage of goods and a further Act of 1894 dealt specifically with this question. 1896 stands out as the year in which the Light Railways Commission was sanctioned by Parliament. This was to increase railway facilities by constructing light railways to add to the existing network and compulsory purchase, together with financial assistance from the State, was mentioned for the first time.

The end of the nineteenth century saw the tendency towards amalgamation continue. This was accompanied by additional facilities being offered by the railway companies, embracing more non-stop trains and improved types of passenger coaches, including corridor carriages, restaurants and buffet cars, and sleeping cars. Efforts were made to keep charges down, and excursions, workmen's fares, weekend and fortnightly and tourist tickets at reduced fares were introduced. In addition, the railway companies established hotels in most large towns and they began to purchase fleets of steamships so they could transport passengers between Britain and the Continental ports and Ireland.

At the same time as these improvements were being made there was a growing motor car population. Railway working expenses became heavier because the increase of the new facilities required the construction of more and better rolling stock with greater haulage power. Improved safety conditions, brakes and signal gear all added to the overhead costs, which began to create severe financial problems.

2.8.3 Suburban growth

The closing years of the nineteenth century were not short of agitation for the improvement of housing and working conditions. Charles Booth[38] was advocating the provision of a large and complete scheme of underground and overhead railways extending beyond the existing boundaries to wherever the population might go. The Royal Commission on London Traffic Report gives figures for 1901. Something like 269662649 passengers travelled by local railway, tram and bus in 1881 and the report mentions that by 1901 this figure had risen to 847212335.

Suburbanization had a dramatic impact on the property scene around the City. Artisans moved out of their small cottages. These were taken over by the very poor whose slum property was replaced by warehouses and some offices. Mayfair, Kensington and parts of Chelsea provided homes for the rich but the artisan workers were moving out to North Paddington, Hammersmith, Fulham, West Ham and Edmonton and similar districts. The managerial and middle classes were beginning to invade Highgate, Hampstead, Putney, Wimbledon, Wandsworth, Brixton and beyond. There were physical changes in the environment. The house standing in its own grounds tended to disappear. With few exceptions, flats for the artisans and middle classes were not being built as yet, these people being housed in rows of small neat terraced or semi-detached houses in such districts as Clapham, Brixton and Streatham. Particular districts took on specific class characteristics, Acton and Ealing for instance became very popular with the middle classes when these were connected by tram and tube to Central London.

By 1899, the Great Eastern Railway was running fifteen cheap return trains, including seven from Enfield, under the Cheap Workmens' Fares scheme (at twopence (2d) a time) covering a distance of 21½ miles — or just over 10 miles for a penny. At that time some 6 million twopenny returns and 4 million half-fare passengers were arriving at Liverpool Street Station each year; all of them before 8.00 a.m.[39]

Policing the 1883 Act was the Board of Trade, which found itself involved in a court action where it was decided that no railway company could be forced to reduce fares if by so doing it was running at a loss. The situation in London by this time was desperate. In the central districts commercial expansion forced out residential accommodation to a large extent, and what was left was seriously overcrowded. To relieve the situation, cheap transport to outlying districts was essential and it was indeed fortunate that the era of the electrically-powered tram-car and internal combustion motor-bus made its début.

2.8.4 Tramway expansion

The 1870 Tramways Act allowed private promoters to extend their prime routes but they soon ran into difficulties with the local and road authorities who insisted upon improvements being carried out at the promoters' expense. The legislation, while allowing for the laying of tramlines and the necessary road widening, made the promoters liable to pay compensation to owners of property affected. In some cases the promoters found themselves liable to find the funds necessary to purchase house and shop properties at full value as well as meeting claims for disturbance. There is no doubt that the compulsory purchase provision of the 1870 Act prevented a great many tramway lines being laid. However in 1888, the London United Tramways Company commenced an extensive system in the west and south-west suburbs, the first of the lines being from Shepherds Bush Underground Railway terminus and Hammersmith, to Kew Bridge and Acton. This was followed by lines to Hounslow, Uxbridge, Twickenham, Hampton, Teddington, Kingston-upon-Thames and even Surbiton. There were local routes commencing at Kingston which connected with Merton via New Malden, Raynes Park and Wimbledon. Almost all of these lines were new, only a small part in the Hammersmith area being converted from the original horse/tramway system.

Another private company, the Metropolitan Electric Tramways, operated in the north and north-east of London, going as far as Edgware, Edmonton, Waltham Cross, Wood Green,

North Finchley and Golders Green. It also branched west from Harlesden to Acton. In the east this was extended with routes to East Ham, Ilford, Barking, West Ham, Walthamstow and Luton. South of the river the company went as far as Erith, Dartford and Bexley. The area around Croydon was developed by the Corporation itself, and included Thornton Heath, Selhurst, Anerley and Addiscombe. The South Metropolitan Electric Tramways and Lighting Company linked Croydon with Tooting and Sutton and went as far as Mitcham. It also opened up branches to Crystal Palace and Penge. In the London County Council area the Council itself laid many lines of tramway.

Although the first motor buses appeared in 1899, they hardly competed with the trams because of their tendency to break down, their lack of comfort and inefficiency. By 1908 a number of the smaller firms were amalgamated and the well-known London General Omnibus Company was running as many as 899 buses. This company came into its own when it had built for it a new bus known as the 30 hp B type, which was probably the first efficient and reliable vehicle of its type. It became a serious rival to the tramway system, since it did not depend upon the tramway lines which governed the route of the trams, and yet could compete at the same time with the tram routes themselves.

During the ten years leading up to the First World War the activity of the tramway and bus system caused a loss to the steam railway system. The railway companies were not slow to open new stations in close proximity to London, and one railway company, the District, commenced the first above-ground electric railway operation. In 1903 other companies opened 'suburban lines' mainly powered by electricity. The tramway development pattern between 1901 and 1939 confirms the competition between bus and tram and the new suburban railway stations that were opened from 1901 to 1940. Even before 1914 the trams of the central area district were suffering from the competition following the introduction of the 30 hp LGOC B type bus.

2.8.5 The underground

The introduction of electric traction enabled deep underground lines to be built. The construction costs were enormous and the return on capital negligible. However, the capital already involved in the underground (or 'tube') venture was too great to risk further income loss and there was no real alternative but to extend further outwards in order to generate new business and in fact 34 new stations were built in the London area by 1914.

2.8.6 Population, development and transport

The relationship between population, development and transport calls for the compilation of statistical information, from which general conclusions may be drawn. In the case of London, a study of the effect of changes has been documented in great detail by Alan A. Jackson.[40] The population figures show an increase from 1901 to 1921 (roughly double) and although the same rate did not occur between 1931 and 1939 in every area, selected areas do show substantial increases. This must have had a considerable impact on the transport necessary to serve these areas.

The era that began with the first steps towards the mechanization of labour and ended with the First World War produced a dramatic effect on the built environment, its owners and inhabitants. A coincidence of circumstances fitted together like pieces in a jigsaw puzzle. The raw materials bought on preferential terms from colonies throughout the British Empire supplied the new machines with the fuel to power them, which by happy accident was underneath their feet. Competition from other countries had been negligible and the finished manufacturies were sold to the very countries that supplied the raw materials in the first place. There were, however, signs that the United Kingdom could not maintain its near monopoly trading status: other countries emerged as competitors, especially in the manufacture of steel and cotton goods. This and the introduction of refrigerated ships contributed to the decline of the United Kingdom's erstwhile world wide trading pre-eminence from the 1870s onwards.

Factories sprang up close to the fuel supplies, such as by water (sometimes a source of power, as in the case of mills in Yorkshire), or where adequate road, canal and later railway facilities were readily available. Almost all these locations existed in established towns and cities and as the factories increased in size and number so did the demand for workers' accommodation. For the most part, the complete absence of any control on development by entrepreneurs motivated by profit and little else resulted in the creation of environments, the best of which were barely tolerable and the worst almost a crime against civilized humanity.

Throughout the ages mankind has witnessed rebellion and revolution arising from oppression territorially or politically engendered, or merely from hunger, (as for example the French Revolution (1789) which began after crop failure). If not a revolution, then the evolution of the social contract with which Karl Marx is associated came as a result of the writings of Engels who had been sent by his family firm to Manchester to help run their factory there. He used reports of the 'Government committees on the living and working conditions of the labouring classes of Britain', to which he added his own personal observations.

Property is perhaps the most visible of all those items that make up the sum of human requirements. The abuse of child and female labour that characterized the earlier part of the era occurred in the depths of the earth or in those 'dark Satanic mills' of which Blake wrote, which were invisible to those on the outside. However, the slum tenements to which the unfortunates returned after twelve hours or more slave labour were only too painfully on permanent exhibition and drew attention to the living and working conditions of the inhabitants.

The social history of the era records the various Acts of Parliament passed to remedy the worst of the conditions of work; but there is little evidence of improved housing for which there was no mandatory legislation and clearance of slums occurred mainly when the sites on which they stood were found to have potential for commercial redevelopment.

Both national and local taxation emerged as factors affecting ownership of property. Firstly the land tax and then the thin end of the wedge of income tax affected investment income and more especially the duties payable on death led eventually to the break-up of estates. Towards the end of the era local authorities were forced to spend increasingly large sums on facilities necessary to accommodate the enlarged populations, a direct result of the concentration of people in towns and cities due to the Industrial Revolution. This led to the collection and consequent spending of the rates, the amount of which, in time, became the equivalent of the annual value of the properties within the boundaries of the authority.

While not underestimating the importance of the emergence and development of mechanized transport, its main effect on property was to determine its location. Most of the factories and the houses that complemented them were in locations chosen for easy access to ports, raw materials and power or the fuel to generate it. Roads, canals and railways tended to service already existing centres rather than create new ones. The exception was the introduction of local passenger transport that encouraged suburban development. It allowed workers to move from inner city older properties to new suburban housing which was generally of better quality. This trend occurred as a result of purely commercial considerations with minimal Government interference and with no control on prices or rents until the outbreak of the 1914 war, when rent control and security of tenure for tenants of low rented accommodation were introduced to prevent profiteering by landlords who intended to evict sitting tenants so as to re-let to munition workers earning much higher wages. The legislation was intended to be temporary but subsequent events proved otherwise.

Towards the end of the era two important events occurred that were to have consequences in the future: the Town Planning Act and the Finance Act of Lloyd George's Liberal administration. Although permissive only, the Town Planning Act paved the way to a new vista that was to unfold in coming decades. The Finance Act, however, had the greater significance in that it was a direct attack on property owners and the forerunner of the legislation that did not occur until almost a half century later!

In retrospect, it may be held that running alongside the growth of industrialization (with all its benefits) ran the social evils that society at first permitted and then only partially eradicated by the end of the era. Of those evils, that of the built environment showed the angriest of sores and it was left to succeeding generations to attempt their healing.

REFERENCES

1. Poor Law Commissioners, Report on an inquiry into the Sanitary Condition of the Labouring Population of Great Britain, p. 41.
2. Godwin, G. (1972) *Town Swamps and Social Bridges*, Leicester University Press pp. 67—68.
3. Select Committee on Health of Towns.
4. Royal Commission on State of Large Towns, etc. 1st Report, pp. 150—151.
5. Farr, W. (1885) in *Vital Statistics* (ed. N. A. Humphreys), Stanford University Press, p. 467.
6. Farr, W. (1885) in *Vital Statistics* (ed. N. A. Humphreys), Stanford University Press, p. 200.
7. Farr, W. (1885) in *Vital Statistics* (ed. N. A. Humphreys), Stanford University Press, pp. 60—64.
8. Richardson, B. W. (1887) *The Health of Nations. A Review of the Works of Edwin Chadwick*, Vol. II, Longman pp. 317—318.
9. Robinson, G. T. (1871—72) On town dwellings for the working classes, *Trans Manchester Statistical Society*, p. 76.
10. Robinson, G. T. (1871—72) On town dwellings for the working classes, *Trans Manchester Statistical Society*, p. 81—82.
11. Child, G. W. (1878) How best to overcome the difficulties of overcrowding among the necessitous classes, *Trans National Association for Promotion of Soc. Science*, p. 494.
12. 14 & 15 Victoria c. 34.
13. Postage, R. W. *The Builders' History*, p. 197.
14. Dearle, N. B. (1908) *Problems of Unemployment in the London Building Trades*, Dent, London, p. 19.
15. Howarth, E. G. and Wilson, M. (1907) *West Ham: A Study in Social & Industrial Problems*, Dent, London, p. 12.
16. 53 & 54 Victoria c. 59.
17. 54 & 55 Victoria c. 76.
18. Hamer, J. What are the best means, legislative and otherwise of securing those improvements in the dwellings of the poor which are essential to the welfare of the community? *Trans National Association for the Promotion of Social Science*, p. 468.
19. Select Committee on Artisans' and Labourers' Dwellings Improvement, Interim Report, p. 134.
20. Costelloe, B. F. C. (1898—99) The housing problem, *Trans Manchester Stat. Socy*, p. 53.
21. Select Committee on Artisans' and Labourers' Dwellings.
22. 6 Henry VI, c. 5.
23. 7 & 8 George IV, c. lxxvi.

24. 1 & 2 William IV, c. xlv.
25. Rumsey, H. W. (1871) *On a Progressive Physical Degeneracy of Race in the Town Populations of Great Britain.*
26. Horsfall, T. C. (1901) *The Relation of Town Planning to the National Life*, p. 13.
27. Bright, T. (1909) The development of building estates. *The Times*, 1 September.
28. Social Science Association (1909) *Transactions of the National Association for Promotion of Social Science*, p. 454.
29. Nettlefold, J. S. *Slum Reform and Town Planning. The Garden City Idea applied to existing Cities and their Suburbs*, p. 2.
30. Unwin, G. (1924) *Samuel Oldknow and the Arkwrights.* Manchester University Press, pp. 170, 175.
31. Salt, T. (1853) *Manchester Guardian*, 21 September.
32. Reckitt, J. (1908) *Garden Cities and Town Planning, New Series*, III, 97.
33. Ebenezer, H. (1946) *Garden Cities of Tomorrow.* Faber and Faber, London.
34. Barnett, H. *Cottages with Gardens for Landowners.* Hampstead Tenants Ltd, p. 6.
35. 6 Edward VII c. cxcii.
36. Smith, A. (1766) *Inquiry into the Nature and Causes of the Wealth of Nations*, Book II, Part II, Chapter 5.
37. Paine, T. (first published 1791) *The Rights of Man.* Published in 1906 by Dent, London.
38. Booth, C. (1901) *The Improved Means of Locomotion as a First Step towards the Cure of the Housing Difficulties of London.* Walworth, p. 13.
39. Morris, O. J. (1953) *Fares Please: The Story of London's Road Transport System.* Allan, London.
40. Jackson, A. A. (1973) *Semi-Detached London.* Allen and Unwin Ltd.

FURTHER READING

Burke, G. (1980) Town planning and the surveyor. *Estates Gazette*, London.

Burke, G. (1971) *Towns in the Making.* Edward Arnold, London.

Burnett, J. (1978) *A Social History of Housing.* David and Charles, Newton Abbott.

Chaloner, W.H. and Musson, A.E. (1963) *A Visual History of Modern Britain — Industry and Technology.* Vista, London.

Chamberlain, R. (1983) *The National Trust — The English Country Town.* Webb and Bower, Exeter.

Cherry, G.E.(1974) *The Evolution of British Town Planning.* Leonard Hill, Leighton Buzzard.

Cullingworth, J.B. (1978) *Town and Country Planning in Britain.* 7th edn. Allen and Unwin, London.

Emeny, R. and Wilks, H. (1984) Principles and practice of rating valuation. *Estates Gazette*, London.

Goldsmith, M. (1980) *Politics, Planning and the City.* Hutchinson, London.

Hall, P. (1974) *Urban and Regional Planning.* Penguin, Harmondsworth.

Lloyd, D.W. (1984) *The Making of English Towns.* Victor Gollancz, London.

Offer, A. (1981) *Property and Politics 1870—1914.* Cambridge University Press, Cambridge.

Olsen, D.J. (1982) *Town Planning in London — The Eighteenth and Nineteenth Centuries.* 2nd edn. Yale University Press, London.

Pressat, R. (1973) *Population.* Penguin, Harmondsworth.

Roots, G. and King, N. (1984) *Ryde on Rating.* Third cumulative supplement to 13th edition. Butterworths, London.

Skeffington, A. (1969) *People and Planning.* HMSO, London.

Sutcliffe, A. (1981) *Towards the Planned City.* Blackwell, Oxford.

Widdicombe, T. and Anderson, E. (1976) *Ryde on Rating.* 13th edn. Butterworths, London.

3

Wars and Peace

3.1 INTRODUCTION

The popular battle cry of 1914 was that the conflict with Germany was to be a war to end wars. It proved to be a vain hope since the peace that followed lasted only some 21 years. Like the Thirty Years war of 1618–48 the years 1914–45 can be considered as a single epoch. Each was marked by peace treaties honoured in the breaches rather than in the performance. The period between 1914 and 1945 has been compared in one respect to that from 1793 to 1815. In each case the peace that followed the war proved to be no more than a longer than usual pause between battles.

Changes in the history of property development did not coincide with the dates of successions to the throne or the declaration of wars and the peace treaties that followed. Rather the changes did occur as a result of shifts in political power influenced by a continuation of economic and social factors. These tended to encourage improvement and control of the development of the built environment and required legislation to give them effect. Thus the period between the commencement of the First World War until the peace following the Second can be considered as a single era, mainly because the Government was dominated by one political party even during the Coalition Governments of both wartime periods. The direct attacks on property interests introduced by Lloyd George in 1909 and 1911 failed in the event with his party's loss of power and the subsequent declaration of war in 1914. Nothing like his proposals was to emerge in legislation until after 1947.

Yet it was during the hiatus of domestic, economic and social legislation during both wars that the seeds were sown for the crop of legislation that subsequently occurred. In retrospect the more significant changes can be identified at a time when the number of houses built exceeded the demand, the control of rents and security of tenure introduced for the first time the growth of suburban communities furthering dramatic changes in transport facilities and communications. Others (arguably of some lesser importance) affected the rating system and retail distribution with the growth of 'multiples' the latter especially, combined with the growth in road transport changed the location and function of warehouses.

3.2 PLANNING

During both the First and Second World Wars, the Governments of the day were much concerned with housing and planning for post-war periods. It will be remembered that the 1909 Town Planning Act was a permissive Act. It allowed local authorities to prepare schemes so that private owners could improve their own properties within a town planning scheme, instead of acting on their own initiation. The general provision of each scheme had to be approved by Parliament on the recommendation of the Local Government Board. The total number of schemes authorized to be prepared was no more than 172, covering nearly 300 000 acres in total. Only 13 of these schemes were actually submitted before the second of the Town Planning Acts, the Housing and Town Planning Act of 1919.[1] Before and during the war there had been considerable lobbying, particularly by the National Housing and Town Planning Council, campaigning in favour of new legislation, especially in the provision of working-class housing.

In 1916, the Government set up a Reconstruction Committee which was reorganized in 1917, and from which developed the concept of a new Ministry of Health and a massive housing programme for the working classes. The committee's brief included a study of commercial and industrial policy, and Local Government housing, as result of which, in 1919 it made proposals for a Ministry of Health, a Ministry of Transport, and recommended a new Housing and Town Planning Act to include legislation on the assessment of land values.

The National House Builders' Council submitted a recommendation that 400 000 new working-class houses were needed. The Royal Commission on Housing set up in 1912 to examine the industrial population of Scotland reported in 1917, after five years' investigation. It was this report that highlighted not so much the quantity, but the quality of the new housing that was needed. The Local Government Board, which had been set up under the 1909 Act, had sponsored the Tudor Walters Committee to consider building construction and the provisions of dwellings for the working classes. Their report was submitted in 1918 and its proposals covered density, site planning and house design and recommended amongst other things that in working-class areas, development should not exceed twelve houses to the acre in urban areas and eight to the acre in rural areas. They recommended that each house should have its own garden, large living rooms, a bathroom, an indoor WC, a larder and a cold store. These standards were adopted in the Housing Manual of 1919, and remained as the basic standards until the Dudley report of

July 1944, which was itself superseded by the Parker—Morris report of 1961. The end result of the reporting of the earlier committees was the Housing and Town Planning Act of 1919 which was a considerable extension of State intervention in this field.

The Act of 1919 imposed in Section 1 the duty of every local authority to consider the needs of its area and to prepare and submit to the Local Government Board, a scheme for the exercise of its powers. Neville Chamberlain, when Lord Mayor of Birmingham in 1915—16, admitted that the best choice of site was often thwarted and checked in town planning, 'because we find that what would be best for the community will involve injustice or hardship to individuals.' Section 42 of the Act provided that it was no longer necessary for the local authority to obtain the Local Government Board's permission to prepare or even adopt a town planning scheme. The object of this section of the Act was designed to expedite the preparation of schemes and thus encourage actual development. Section 46 required every local authority with a population of 20 000 or more to prepare a planning scheme for any land either in the course of development or likely to be developed.

There is a little more to the 1919 Act than its predecessor, although the newly formed Ministry of Health had power to take action against a local authority who neglected to prepare a scheme. From the point of view of the planners, the net result of the 'compulsive' power given to the Ministry must be considered a failure, since the subsequent Housing Act of 1923 allowed the period for the preparation of schemes to be extended to January 1929. Those authorities who failed to prepare a planning scheme found that no action was taken against them and in 1932 the law was actually changed to remove compulsion from town planning law.

The total effect of the first two Acts and the recommendations on the files leading up to these Acts could be said to have proved no disincentive to the production of a vast number of new houses built principally for the working classes who, after all, formed the bulk of the population in the years between the wars. It might be felt that these early Acts did not go far enough but it is considered that had there been more teeth in them, they might perhaps have bitten off the hands that laid the bricks. In the words of Gordon E. Cherry:

> 'Town and country planning was an incredibly tender plant and the more decisive statutory process gave it firmer soil. The nutrients were provided by the propaganda bodies, the growth of the professional institute, and shifts in political and social attitudes in favour of more State and local intervention in matters held previously to be of private concern'[22]

The total number of Britons killed on the Western Front alone during the First World War was 743 702. The German experiment to bomb British towns from Zeppelins had failed ingloriously and the state of affairs at the end of the First World War could be compared with that at the end of the Second by exchanging the destruction of men in the former for the destruction of property in the latter. The population in England and Wales increased by 10.9% between 1901 and 1911. Between 1921 and 1931 it had fallen by 5.5%. In 1913 the birth rate was 24.1 per thousand population. By 1939 it had fallen to 14.8 per thousand population with 40% of the population living in seven distinct areas — London, Manchester, Birmingham, West Yorkshire, Glasgow, Merseyside and Tyneside. Greater London had a population of some eight million, which was about one-fifth of the total population of England and Wales.

Development followed the lines of road and rail communications. London, for instance, was developing its underground railway system in all directions and around each station the new housing estates were eagerly snapped up. The total population shifted from region to region, so that some areas showed marked increases while others decreased. The Home Counties, and especially London, revealed the highest growth. Between 1921 and 1937 the increase in the Home Counties was 18%. South Wales recorded a decrease of 9%, while others to show increases were the Midland Counties, by 11%, The West Riding, Nottinghamshire and Derbyshire, by 6%, and surprisingly, the Lowlands of Scotland, by 4%. Added to the vast numbers killed during the war was the tragedy of the influenza epidemic of 1918–19, which resulted in the deaths of more than 150 000 people — the highest death toll since the cholera epidemic of 1849. Once again, the interrelated effects of health, housing and fitness for military service, or the ability to withstand a dose of the 'flu were emerging. Only 36% of the recruits in the years 1917–18 were passed as Grade 1. In the Second World War this figure was 70%.

New techniques in engineering and electronics were to have a marked effect on the distribution of workers in almost every sphere of activity. New machinery had affected the numbers required for the textile industry and agriculture. The employment of women in factories during the war was to persist and free many of them from the domestic service they were compelled to undertake for a living before the war. There was a change in the areas where one found unemployment. Before the war, unemployment in London fluctuated between 4 and 8%, it reached 20% in some parts of the North, Scotland and Wales.

Something like four million dwellings were built between 1919 and 1939, of which approximately 400000 were built by local authorities and an average of 180000 a year by private enterprise. Most of the latter were bought by individuals through the medium of the building societies and the supply of homes outstripped the growth of population.

3.3 POPULATION

The census figures estimate that the total population of the UK was 40831000 in 1911, 42769000 by 1921 and 44795000 by 1931. There were dramatic changes in the population of specific towns and suburbs of London and the South East, due almost entirely to the increased transport facilities. Examples of population increases in certain areas, between 1931 and 1938, are shown in Table 3.1. The population of Greater London, which in 1901 was some 6½ million, had increased to 8¾ million by 1939. Selected extracts from the London County Council statistical records[3,4,5] show the dramatic increase in population, approximately 100%, that occurred in certain areas between 1931 and 1939 (Table 3.2). Table 3.3 illustrates the increase in population of selected districts outside the Greater London Area, between 1921 and 1938.

Table 3.1

District	*Population increase (%)*
Ruislip—Northwood UD	154.6
Bexley MB	135.8
Chislehurst and Sidcup UD	127.4
Potters Bar UD	110.0
Carshalton UD	105.5
Hornchurch UD	92.9
Hayes and Harlington UD	91.3
Harrow UD	89.8
Feltham UD	89.5
Orpington UD	79.1
East Barnet UD	77.0
Epsom and Ewell MB	70.1
Worthing RD	68.8
Rickmansworth UD	62.2

The Town Planning Act in 1925 was especially important as it was the first planning legislation not primarily concerned with housing. It was, in fact, a consolidating Act to encourage local authorities to progress their proposals and, as such, instituted the move towards a comprehensive centralized planning system. The legislation required authorities to submit planning schemes by 1928 but there were still 98 (out of 262) urban authorities with a population of more than 20000 which had not submitted proposals in any form and

Table 3.2

	Population in 1931	*Population in 1939*
Banstead	13089	24480
Bexley	32626	80110
Epsom and Ewell	35228	62690
Harrow	96656	190200
Hayes and Harlington	22969	50040
Ruislip and Northwood	16035	47760
Uxbridge and Wembley	65799	121600

Table 3.3

	Population in 1921	*Population in 1938*
Slough	20285	50620
Billericay	12431	34730
Hornchurch	17489	76000
Welwyn Garden City	901	12150
Crayford	12295	24590
Coulsdon and Purley	23115	55070

some of the others had done so in part only. The date for submissions of obligatory schemes had already been extended and the Local Government Act of 1929 extended it further to 1 January 1934, with power to the Minister to extend the date to 31 December 1938. By 1930, there were still 58 urban authorities which had not submitted proposals. A further Town and Country Planning Act in 1932 extended legislation to cover rural as well as urban areas and incidentally brought into legislative control some nine million acres of England and Wales.

The Housing Act of 1923 provided for compensation for compulsory acquisition and included an authorization to the Minister to preserve the existing character and features of a locality with special architectural or historic interest. This was the first appearance of conservation in planning legislation and Oxford, Canterbury, Exeter and Winchester were among the first to apply its provisions. The Local Government Act of 1929 gave county councils the right to share in joint town planning schemes with local authorities, without the power to initiate the schemes themselves. However, Abercrombie and Unwin, two outstanding figures in the world of town planning were called in and joint town planning advisory committees were constituted in 1920 for Manchester, Chester, North and South Teeside, North and South Tyneside and West Middlesex, amongst others.

Raymond Unwin was appointed technical adviser to the most important of the regional committees, set up by the Minister of Health to consider a plan for Greater London. His first report, which was issued in 1929, was significant in that it dealt with the question of a green belt and control of ribbon development on main roads, and recommended some 62 square miles of additional playing fields and 142 square miles of open space. He also recommended the building of new towns, described as 'satellite communities' at distances of between 12 and 25 miles from Charing Cross, to take the form of industrial garden cities. Unfortunately, by 1931 the downturn in the economy of the country prevented any implementation of these proposals and even the expenses of the committee were curtailed. By 1926, 33 advisory and one executive committee set up under the 1923 Act had been established, and by 1932 the figures were 60 and 48 respectively. Some 30 regional planning schemes in England had been studied, besides three in Scotland. The Minister of Health in 1931 set up another committee to consider all these reports and to estimate what was now required. The committee reported that they saw no case for any special organization, whether national or local, to concern itself with planning, since they thought that Government and local authorities had sufficient machinery already.[6]

3.4 POLITICAL AND OTHER INFLUENCES ON PLANNING

It would seem that, irrespective of the political leanings of the Government of the day, the efforts of those professional town planners of the era were, in the final analysis, able to see only a tiny part of their proposals brought to fruition. Stephen Ward commented on the Town and Country Planning Act of 1932 that 'any interference with the operations of private enterprise was taboo. Public schemes (particularly those that could be labelled as Socialist) were speedily decimated with little regard for long-term effects.' Free enterprise Capitalism, it was argued, would save the day, and wishy-washy Socialism, public control and the like would only make matters worse. He interpreted four distinct species of political philosophy existing at that time. First, the Labour party whose aims were basically to press forward towards the achievement of the Socialist State more or less as outlined by Marx, but by gradual, not revolutionary, means and working through the Parliamentary institutions rather than subversion. Planning, to the Socialist, was a gradual means of controlling the land in order to use it in the interest of the population as a whole, rather than in the interests solely of the landowners. Secondly, there was the planner, not particularly political and supposedly, where planning was concerned, divorced from any party, but seeing a crisis and proposing a rational long-term and co-ordinated policy on the best use of national resources. Thirdly, there were the enlightened Conservatives of whom he cites Neville Chamberlain as the best example.[7] An early advocate of planning in Birmingham where he had been Lord Mayor, he was nevertheless an industrialist and Capitalist. On the one hand, he believed that workers functioned better in pleasant surroundings than otherwise while, on the other, he wanted the towns and countryside to be developed sensibly. The financial sanctions necessary to make planning controls effective bit deeply and painfully into his belief in the Capitalist system. Lastly, there was the die-hard Tory completely opposed to legislation in the private sector.

According to the historian, Arthur Marwick, there was, however, a nucleus in the Tory party supporting town planning. Amongst the personalities was Harold MacMillan, later to become the Minister responsible for Housing in Churchill's Government and subsequently Prime Minister. Sir Ernest Simon, at the Town and Country Planning Summer School in 1937, postulated that

> 'whatever political party we may belong to, we believe that Democracy is the best form of Government, and that in respect of town planning without falling behind Russia and also other countries under dictatorship, I firmly believe that if British Democracy makes up its mind, it can within the next two generations make the cities of England once more, places of beauty in which it is good to be alive.'[8]

The Act of 1932 was a great disappointment to the advocates of town planning outside Parliament. The President of the Town Planning Institute, R. C. Maxwell, was particularly outspoken in this comment on the composition of the standing committee which examined the Bill before it became an Act.

'The absolute ignorance of any idea of town planning which exists in the mind of some Members of the Standing Committee is almost unbelievable -- not understood by people who think their accidental position in the House of Commons, constitutes them as an authority upon any subject. They know nothing of the contents of the former Act, nor the ordinary procedure under legislation of this kind.'

The Act concluded by re-introducing controls that had been eliminated in 1919. Betterment, nominally fixed at 75%, was found impossible to collect because of restrictive clauses in the Act.

Stephen Ward commented that 'town planners saw their Bill wrecked by ignorance and blind pursuit of an out-moded political faith' and there began to grow the idea that the planner should take 'an increased part in the defence of Democracy'. In common with most other town planners, it is difficult to dissociate him from professing Socialist principles, although he says 'The planner, rather than being a truly political animal, has become a dispeller of ignorance and short-sightedness and a purveyor of rational long-term solutions.' Can such a claim be acceptable? However much it may be denied by the planners, is not this attitude consistent with the achievement of the Social State more or less as outlined by Marx but by gradual — not revolutionary — means?

There were further advances in town planning proposals after the 1932 Act. The drift to the South East and the consequent unemployment led to the designation as 'distressed areas' of large parts of the north of England and Wales. There was a Commission for Special Areas (England and Wales) and companies were set up for the provision of trading estates to attract new industries to these distressed areas. These were financed from the Special Areas Fund and a start was made in 1936 at Durham. The two largest were at Treforest in South Wales and Hillington near Glasgow.

3.4.1 Difficulties of implementing the planning proposals

The 1932 Act (Section 3) reconstructed the Greater London

Regional Planning Committee, this no doubt being due to Raymond Unwin's original proposal for a green belt around London, contained in his second report of the Greater London Regional Planning Committee in 1933. The Act followed an intensive public enquiry opposed by the major London landowners, nevertheless in May 1935 the proposals to subject all London to planning control were put into effect. The procedures for finalizing a scheme were so complicated that by 1939 nothing was approved. To add to the frustration of the planners it was found that the highway and housing departments were not subject to the same planning control regulations as private developers, and these departments were uncooperative with the planners. Furthermore, the number of planning applications which arose necessitated a huge increase in the numbers of planning staff. London was not alone in this experience.

There was one further Act which had some importance. The Restriction of Road Ribbon Development Act 1935 provided for planning authorities to adopt a standard width for any road in their areas, also the consent of the highway authority was required for access or development within 220 feet of the middle of a classified road. In practice, this resulted in service roads parallel with the main highways, but its real purpose, not immediately apparent, was to break up linear development and secure grouping in depth.

It is interesting to relate the reminiscences of Alan Holland, a director of Wates Built Homes Limited, who in May 1978 gave an outline of his firm's activities. Wates was one of the larger firms of house builders whose contemporaries included New Ideal Homesteads, Wimpeys, Laings, Taylor Woodrow and others who were responsible for building many of the smaller houses in the suburbs of London, and their experiences were no doubt shared by firms such as Leech of Newcastle and others operating in similar provincial towns. During this inter-war period, Wates built some 30 000 houses almost exclusively in one area embracing Sidcup, Catford, Lewisham, New Malden, Streatham and Croydon.

The market Wates sought to capture was the bus-driver, the postman, the policeman and the pensionable worker, earning at that time anything from £3 to £4 per week. In 1934, the Company was able to produce at a selling price of

£295, a terraced house having two bedrooms, two living rooms, a kitchen and a bathroom, with small gardens in the front and rear. Those with three bedrooms were sold at £400. A purchaser could secure a house with as little as £5 deposit with the whole of the balance on mortgage repayable over 21 years. This worked out at about a half-crown (12½p) per week for every £100 borrowed, and anyone buying a £400 house could reckon on a total outgoing of no more than 10s (50p) per week excluding rates. The significance of this weekly payment was that it was roughly one-sixth of the purchaser's take-home pay, because at a wage of £3 per week there was no income tax payable.

Wates' philosophy was that by concentrating the development it undertook in one area only, it could keep a permanent work force and need not rely upon casual labour. Moreover, the company recruited its work force from the locality in which it developed and generally took on its payroll only those who were within cycling distance of the area. The profit on each house was a determining factor in the price that was asked. As little as £30 per house was deemed sufficient, because that represented the price that was paid for the plot, and Wates worked on the principle that if it could get the money out of a completed house quickly by selling cheaply, it would buy the next plot that much quicker. In contrast, Alan Holland says that in today's market Wates must sell ten houses to buy one plot, and he gives the reasons which have brought about this differential as higher building costs, increased interest rates, coupled with much longer time delays (which add to the interest charges) plus taxation on sales profits.

The quality of houses produced by the bigger builders was better than most of those built by the smaller and sometimes one-man builder speculators. Some of the smaller builders produced houses which could only be described as jerry-built. In 1939 Norman Wates, together with the directors of some of the other larger firms, got together to form an association which laid down minimum standards for construction and amenities and later became the National House Builders' Council. It is interesting to compare the proportion of the 1930s' take-home wage (after tax deduction) to the cost of accommodation in weekly terms with that of the mid-1970s and to consider to what extent price is influenced by planning and tax, since it is difficult to remain content with the assumption that inflation on property has exceeded inflation on wages to produce such anomalies as now exist.

3.5 REPORTS PREPARED DURING WORLD WAR II AND THE SUBSEQUENT LEGISLATION

The most significant events in planning took place as a result of reports commissioned prior to and during the Second World War. The relationship that existed between these reports, the Civil Service and the politicians of both parties who were responsible for the legislation that followed after the war is also significant. To quote Cullingworth:

> 'Would a committee of officials in any other time so interpret their terms of reference (as did the Whiskard Committee in 1945) as to include a rejection of the policy framework within which they were asked to report? Ministers, of course, played their part, and without Reith, W. S. Morrison and Silkin, very different proposals may have emerged. Without Anderson, Woolton, Jowitt and Herbert Morrison, the final outcome might have been different. But pride of place in the story goes to now insufficiently remembered officials such as Harrison, Whiskard, Hull, Schaffer and (in the Treasury) Gilbert. It seems unlikely that they would have had such an influence on policy in normal peacetime circumstances. Yet, the ending of the war and the return of single-party government appears to have had remarkably little effect. The foundations laid by the reconstruction secretariat and the planning ministry were unshaken. Whether the return of a Conservative Government in 1945 would have made much difference is a matter for speculation; on the evidence of the history of the regime of W. S. Morrison it seems unlikely. (The role of MacMillan in relation to the continuation of new towns policy in the early "fifties" reinforces this view – as a forthcoming volume will demonstrate). Given a practical problem and strong advice from officials, ministers can be seen to defend, develop and propose

policies which simple politics analysis would find surprising.'[9]

3.5.1 The Barlow report

In 1937, Sir Montague Barlow was appointed chairman of a Royal Commission to consider the geographical distribution of the industrial population of Great Britain. The terms of reference included the causes which influenced the geographical distribution of this population, the probable direction of any change in the future, the consideration of the social, economic and strategical disadvantages arising from the concentration of industries in particular areas and, furthermore, the Commission was to report on what remedial measures should be taken in the national interest. The Commission, in its report, produced additional technical evidence and views on the national and urban situation of the inter-war years and, in particular, commented on the state of overcrowded cities. Its report, completed in August 1939, was presented in Parliament in January 1940. It made three main recommendations:

1. A central planning authority, national in scope and character, should be created.
2. Congested urban areas should be redeveloped for residential purposes excluding industry.
3. New industrial development should be redistributed on a regional basis avoiding concentration in any one region.

Particular emphasis was made on the social, economic and strategical problems which, the report advised, demanded immediate attention to arrest the continued drift of the industrial population to London and the Home Counties. Three of the members of the Commission, including Professor Abercrombie, wanted a new Ministry formed to deal with this aspect.

3.5.2 The Uthwatt report

Under the Chairmanship of Mr Justice Uthwatt, an expert committee on compensation and betterment was appointed in January 1941. It was set up 'to make an objective analysis of the subject of the payment of compensation and recovery of betterment in respect of public control of the use of land'. In the words of Gordon E. Cherry, 'in so doing, it was tackling a planning problem which previous legislation had not resolved and which practical experience had shown, particularly during the 1930s, was gravely prejudicial to the interests of good planning.'[10] But one of the problems the committee was asked to advise upon was the steps to be taken before the end of the war to prevent reconstruction being prejudiced, as it was thought that blitzed land required new planning powers. The Commission made an interim report in July 1941, suggesting that interim control of development should be extended throughout the country to prevent work being done which would prejudice reconstruction, and further it recommended that in certain cases special reconstruction areas should be defined to enable them to be replanned comprehensively. The final report of the Commission was made in September 1942. In view of the legislation that followed, the recommendations are remarkable in their similarity to those which subsequently became enacted.

1. On payment of fair compensation the State should be vested with the rights of development in all land outside built-up areas.
2. There should be a periodic levy on increases in annual site values.
3. In developed areas there should be compulsory purchase of war-damaged sites and also land outside developed areas in addition to war-damaged sites, to provide accommodation for displaced persons.

3.5.3 The Scott report

Lord Justice Scott was appointed chairman of the Land Utilization in Rural Areas Committee (1941). The terms of reference of this committee were

> 'to consider the conditions which should govern building and other constructional development in country areas consistent with the maintenance of agriculture and, in

particular, the factors affecting the location of industry, having regard to economic operations, part-time and seasonal employment, the well-being of rural communities and the preservation of rural amenities.'

The report was presented in August 1942 and its recommendations included the establishment of national parks and improved access to the countryside, plus an extension of central planning control over rural development. In retrospect, this report lacked the sharply defined proposals that were to be found in other reports of the same era. Nevertheless, planning law eventually took many of its recommendations into account.

3.5.4 The Ministry of Works and Planning

In September 1940, the old Office of Works had changed its name to the Ministry of Works and Buildings and that most dynamic of administrators, Sir John Reith, became the first Minister, bringing in, as his advisers, professional planners of the calibre of Pepler and Dower; he also formed a special reconstruction group. It was he who was responsible for the Uthwatt and Scott committees. Although he was ultimately dismissed by Churchill, who found him a little too enthusiastic, he had time enough to establish the idea of a central planning authority and in 1943 became the Minister of Town and Country Planning. This Ministry was responsible for England and Wales; Scotland retained its powers through the Department of Health for Scotland. Industrial location came under the aegis of the Board of Trade, and the Ministry of Health became responsible for housing in England and Wales. The Town and Country Planning Act of 1943 extended planning control to all land in the country not already covered by a scheme or a resolution. There was a further Act in 1944, which gave new powers to local authorities; for the first time they were able to buy land to deal with areas of extensive war damage, bad layout and obsolete development — obviously including blitzed and blighted land. This 1944 Act was a particularly important one, since it gave a local authority unlimited powers for almost any development project.

3.5.5 The White Paper on control of land use, June 1944

This White Paper is perhaps the most amazing document produced in the war years. It marks a coalition Government's acceptance of what needed to be done after the war. The first paragraph reads:

> 'Provision for the right use of land, in accordance with a considered policy, is an essential requirement of the Government's programme of post-war reconstruction. New houses, whether of permanent or emergency construction; the new layout of areas devastated by enemy action or blighted by reason of age or bad living conditions; the new schools which will be required under the Education Bill, now before Parliament and under the Scottish Education Bill which it is hoped to introduce later this Session; the balanced distribution of industry which the Government's recently published proposals for maintaining active employment envisage; the requirements of sound nutrition and of a healthy and well-balanced agriculture; the preservation of land for national parks and forest, and the assurance to the people of enjoyment of the sea and countryside in times of leisure; a new and safer highway system, better adapted to modern industrial and other needs; the proper provision of air fields – all these related parts of a single reconstruction programme involve the use of land, and it is essential that their various claims on land should be harmonized as to ensure, for the people of this country, the greatest possible measure of individual well-being and national prosperity. The achievement of this aim is an interest of all sections of the community, both in this and succeeding generations. The Government desire to make that achievement possible.'

If land was to be developed, then consent would be given only on payment of a betterment charge at the rate of 80% of increase in the value of the land. This was to be counterbalanced by compensation for refusal of permission to develop. Both payment and compensation were to be dealt with by a special Land Commission. The war had devastated large areas of cities and towns which needed not just partial rebuilding but replanning and rebuilding on a comprehensive basis. Furthermore, there had been some six and a half years without building of any consequence being undertaken and the surplus housing and industrial office and shop supply that existed before the war had been eroded.

Before the end of the war almost every responsible politician was making speeches exhorting everyone to work harder for export and the rehabilitation of the economy. There was obviously going to be a need for redevelopment on a larger scale and in a shorter time than had even been contemplated before. The war-time battle, once won, would merely set the date for the peace-time battle that would turn destruction into reconstruction.

The Coalition Government contained elements of both the Conservative and Labour parties. The Labour party, together with the planners, saw in the White Paper proposals that, once enacted, would be a permanent part of English Law. The Conservatives, with an eye to the General Election which was bound to take place immediately after the war, must have been content to allow the necessary measures to hasten reconstruction and if they were not vocal on the subject, their subsequent action in reversing the betterment proposals in the 1947 Act reflect their true intention. D. N. Chester, who, with J. E. Meade, was in the economic section of the Cabinet Office in 1945, had this to say about the proposals for a betterment levy (see also p.5):

> 'The State leaves the actual ownership of the land in private hands but takes away the profit motive, the mainspring of private enterprise, by nationalising the development value. If I were labelling the scheme, I would call it "bastard Tory Reform" . . . We feel that the scheme should either go a little further in the way of socialization or should not go so far as to take the profit incentive out of private ownership. Any in-between system is likely to get the worst of both worlds.'[11]

Meade wanted to deal with the matter on a different basis altogether, by raising funds for the payment of compensation not by a charge on development, but by some system of general taxation, either on the increment of land values or on the Schedule 'A' value of all real property. He suggested that the collection of such a tax would have the advantage of simplicity, but even so he commented that it still left a very formidable apparatus of planning control confronting the unfortunate developer. 'I cannot help wondering,' he said, 'how vigorous private enterprise will remain in face of structures of this kind. How much better from the particular point of view of employment policy that both landlord and developer were left with an incentive to develop and were simply concerned to obtain the yea or nay of the planning authority.'

3.6 RENT CONTROL AND SECURITY OF TENURE

3.6.1 Background

The control of rents and with it the security of the tenants' continued occupation has a history extending from the First World War. Originally introduced in 1915 as an emergency measure and moreover intended to be of a temporary nature only, both control and security have persisted in varying degrees no matter which political party has been in power.

Whilst in the main the Conservative party had attempted gradually to decontrol and deny security of tenure in the period between wars, the outbreak of hostilities once again in 1939 reversed this trend as a matter of emergency. Subsequent to the end of the 1939—45 war, considerations of emergency gave place to those of political philosophy. In short, Labour politicians wanted control of rented property, Conservatives a free market. Before discussing the effects of these two opposing philosophies, a brief history of the more important measures that Parliament enacted must be mentioned.

3.6.2 Control in the residential sector

Until 1915, the established pattern of landlord and tenant relationship was governed entirely by contract between the parties. From the days of the Norman conquest, no State interference had occurred either by custom or law. In December 1915, Parliament passed the Increase of Rent and Mortgage (War Restrictions) Act. The preamble to the Act stated its purpose as being 'To restrict, in connection with the present War, the increase of the rent of small dwelling houses.'

Prior to the First World War, house building had slowed down due to lack of demand. The onset of hostilities, however, stimulated demand in areas where armaments and other products necessary for the war effort were manufactured. As work forces were augmented, landlords of small houses, completely unrestricted by law, were able to evict long-standing tenants and re-let at increased rents to the highest bidders. To prevent the continuance of this abuse the Government intended to bring in the control measure for 'The duration of the present War and for the period of six months thereafter and no longer'. Houses affected were within certain standard rental and rateable value limits. The rateable limits were £35 in the Metropolitan Police District or the City of London, £30 in Scotland and £26 elsewhere. The Act further imposed restrictions on landlords and mortgagees who could now no longer evict tenants or raise rents or interest except to cover increased rates or a percentage of the cost of improvements.

The Act was amended in 1917 by the Courts (Emergency Powers) Act, and it was again amended in 1918 by the Increase of Rent and Mortgage (War Restrictions) (Amended) Act. As a result of the Hunter Committee Report[12] a further Act followed in 1919 — The Increase of Rent and Mortgage Interest (Restrictions) Act. This repealed the 1918 Act but the general effect of these Acts, together with a 1920 Act known as The Increase of Rent and Mortgage Interest (Restrictions) Act, which raised the previous rateable limits to £105, £90 and £78 respectively, was to perpetuate the main tenets of the 1915 Act until 1968.

The 1920 Act which consolidated the earlier legislation followed the Salisbury Committee Report of 1919.[13] Apart from raising the standard rental or rateable value limits, increases in rent of up to 40% were permitted while mortgage interest was allowed to rise by 1%, up to a maximum of 6½%. Leave of the Court had to be obtained before distress could be levied or possession recovered for non payment of rent and the Courts were given wide discretion in this respect. Section 13 of the Act (for the first time) brought into control business and professional property, although this provision was meant to operate for only one year. The other provisions were intended to operate for only three years but a Continuance Act of 1923[14] extended the main provisions of the

Act until 31 July 1923. A further Act continued these until 24 June 1925 and The Rent and Mortgage Interest (Restrictions Continuation) Act 1925 extended the provisions until 25 December 1927. Between 1927 and 1932, the Expiring Laws Continuance Act continued the provisions until 25 December 1933. The Rent and Mortgage Interest (Amendment) Act 1933 of Rent and Mortgage Interest (Restrictions) Act of 1938 extended them until 24 June 1942. After the end of the First World War, the greatly increased cost of maintenance and repair forced the Government to allow increases of 10% on 1914 standard rents and in 1920 rises of 40% were allowed. These additional percentages were only permitted, however, provided the premises were kept in a reasonable state of repair.

The 1919 Act[15] extended control to cover all houses of which neither the standard rent nor the net rateable value exceeded £70 in London and £52 elsewhere. In 1920, these figures were altered to £105 in London and £78 elsewhere. The result of this was that all except very large houses were made subject to control. The 1919 Act was extended in 1920[16] to protect not only the statutory tenant but also his widow or any relative resident in his house for six months or more before the time of his death. The statutory tenancy could pass only once, however, and on the death of the second statutory tenant the landlord could resume possession.

Immediately following the end of the war there was a boom in the economy, but this soon ended and, by 1923, the country was in a sharp economic depression. An Act in 1923[17] provided that any house, which became vacant or where the sitting tenant accepted a lease of two years or more, became automatically decontrolled. Some ten years later in 1933 the Rent and Mortgage Interest Restrictions (Amendment) Act altered the position once again. Under this Act, controlled houses were divided into three main groups. The first group, where the rateable value was above £45 in London and £35 elsewhere, became immediately decontrolled. The second group below these values but with a rateable value of at least £20 in London and £13 elsewhere became decontrolled on becoming vacant, while those below these rateable values remained controlled. In 1938 the second of these groups was, in turn, sub-divided; the section consisting of houses with a rateable value above £35 in London and £20 elswhere was immediately decontrolled but the lower section remained permanently controlled.

According to the Ridley Committee[18] the number of decontrolled houses in 1945 amounted to 4.5 million. A coincidental 4.5 million houses built since 1919 were also outside control and it was estimated that 3 million of the latter were in private ownership and mainly owner-occupied while the other 1.5 million were owned by local authorities. Out of a total of some 13 million houses and flats, only about 4 million were subject to control; these were mainly owned by private landlords and had rateable values not exceeding £35 in London and £20 elsewhere. Thus there existed side by side similar houses, some controlled and some decontrolled, with the rents of the controlled houses being in most cases 20 to 30% lower than those decontrolled.

Before the outbreak of war on 3 September 1939, the Rent and Mortgage Restrictions Act 1939 came into force. All houses not subject to the old control and with rateable values of not more than £100 in London and £75 elsewhere were made subject to control, the standard rents being those paid at the date of the Act. New houses and those which had not yet been let were to have the rent at their first bona fide unfurnished letting treated as standard rent. Excluded from a Rent Act for the first time in this Act were local authority dwellings. Unlike the 1915 Act, which was intended as a temporary measure, the 1939 Act made no reference to its intended demise after hostilities ceased and the future of rent control as a permanent fact of life was thereby assured!

The gradual decontrol of housing rents that marked the interim years encouraged considerable investment in new housing. Whilst many of those developed were intended for sale, investors would often purchase a complete estate and let rather than sell to the occupiers. Thus the Government, through the local authorities, was relieved of the necessity of providing houses to let and private enterprise, whilst clearly profiting from the opportunity, fulfilled a necessary social function. Newly-built houses, whether for sale or to let, constituted a completely free market. The spin-off of the shopping parades servicing the new estates entailed expenditure of minor importances compared with that of the houses. With some exceptions, factories and warehouses were mainly owner-occupied, as were office buildings in the principal commercial locations.

In commenting on conditions after 1945 the continued rent control will be shown to have had a dramatic effect on the entire economy that could hardly have been envisaged at its inception.

3.7 TAXATION

The standard rate of income tax at the turn of the century was 1s in the £1 and it remained at around this level until 1914, only increasing to 1s 2d; however it increased progressively during the war years to a maximum of 6s 0d. By 1922 it had come down to 5s 0d and it dropped in successive years to 4s 0d in 1930. From then on until 1939 it varied only by 6d. The advent of the Second World War saw it increase yearly from 7s 0d to 10s 0d. Apart from these variations in the rate of tax no significant changes took place; however there were important alterations in the area of rating.

Section 21 of The Union Assessment Committee Act of 1862 required new valuation lists to be made every five years. During the quinquennium the valuation list could only be altered 'if in the course of any year the value of any hereditament was increased by the addition thereto or the erection thereof, of any building or is from any cause increased or reduced in value.' The Rating and Valuation Act of 1925 changed the areas, authorities and tribunals responsible for rating assessments and radically revised assessment itself. It set up a Central Valuation Committee which was 'to give information and make representations to the Minister of Health on the working of the Act'. In particular it reiterated the need for quinquennial revaluations and new valuations were, in fact, made in 1928, 1929 and in 1934. The list due in 1939 was postponed for an enquiry into possible hardship.

The annual cost of rates to the occupiers of houses, business premises, firms, etc., often comes close to either the rent or rental value. In times of economic recession this has been examined at Government level to prevent hardship to citizens or odium falling on whichever party happens to be in power. The Rate Reliefs Act 1928—29 is a particular example. The depressed state of both agriculture and industry caused relief from rates to be given to the extent of 75% to industrial hereditaments, mines, factories, and workshops used primarily for industrial purposes. Similar relief was given to freight transport hereditaments which included premises used for railways, docks and canals.

This era (unlike that which followed the Second World War) witnessed how the power of central Government was used to maintain the low level of the rate burden. In following chapters it will be seen how the system can be used as an economic regulator, along with taxation on corporate, individual incomes and profits, taxes on capital and control of rents both in the residential and commercial sectors.

3.8 TRANSPORT

The inter-war period witnessed two major changes in the methods of transportation which were destined to have a marked effect on urban development — the motor car and the underground train. Both encouraged suburban development, and the commercial version of the motor car — the lorry — proved to be such a rival to the railway system that it threatened its very existence and to this day continues to operate as an economic embarrassment to the State.

3.8.1 Railways after World War I

When the First World War broke out the Government took over the general control of the railways. They left the working management in the hands of the staff of the companies, whose role was, of necessity, primarily transporting troops and military equipment. Very little new rolling stock was added or improvements carried out during the period 1914—18, after which it was thought that restoring the railway system to full efficiency could only be economically achieved through amalgamations. These had previously been discouraged or forbidden by the Government. In 1921 the 121 railways of Great Britain were grouped into four great companies, namely The London and North Eastern, The London Midlands and Scottish, The Great Western and The Southern. The amalgamations were to take effect from the 1 January 1923.

The Railways Road Tribunal was established to fix rates,

fares and other terms of travel, and it was thought that the amalgamations would result in a better deal for the companies, the public and the railway employees. In fact, all four companies were faced with declining receipts, and continuous friction with their employees. Private motoring, independent of timetables and able to convey people and goods from door to door, was the more serious threat to the railways since the road haulage companies, unlike the railway companies were not 'common carriers.' Because haulage companies could select their loads, they carried the lighter and more valuable produce for which the railway companies would have been entitled to charge fairly high rates, the heavier and cheaper loads were left to the railways.

The Road and Rail Traffic Act 1933 tried to alleviate this situation to some extent by allowing the railway companies to negotiate agreed charges allowing some reduction from the standard rates laid down by the Railway Rates Tribunal. Negotiations started in 1938 with the railway companies, the Minister of Transport and the Transport Advisory Council, were designed to give the railways a chance to compete with the road hauliers, but these negotiations came to nothing owing to the outbreak of war in September 1939. Once again the whole of the railway system passed into the hand of the Government, although as in the 1914–18 War, the day-to-day management continued to be exercised by the railway staff.

In America and elsewhere, goods were in plentiful supply but money was not and many workers had to suffer a reduction in wages. With the exception of food, prices generally were drastically reduced, including the price of property of all descriptions. To give an example, the very first transaction carried out by the author was the sale, in the summer of 1933, of a country house (Ascot Lodge, Bracknell) standing in some four acres of ground with seven bedrooms, three bathrooms, four reception rooms, servants' quarters, garage, stabling and other ancillary amenities. This property, which would probably fetch around £150 000 today, was sold at what was then considered a reasonable price of £4750!

Inspections of many properties in the Home Counties necessitated the author making car journeys via the suburbs of London. New estates of small houses, built in the main by well-known house building firms such as Wates, New Ideal

Homesteads Ltd, Nash, Wimpey, Taylor Woodrow, Crouch, Berg and others were continually seen to be springing up around. A freehold house with both a front and back garden and containing three bedrooms, two reception rooms, kitchen and bathroom, could be bought for around £295 with as little as a £5 deposit, with the balance available on mortgage. Interest and repayments at that time were in the region of a half-crown (12½p) per week for every £100 borrowed.

Additionally in the 1930s, several blocks of flats were being built purely for letting purposes, both in inner London and the suburbs. The principal districts where flat development took place were in Bayswater, starting from Edgware Road near Marble Arch, in Queens Road (later renamed Queensway), Bayswater and in various streets off the Bayswater Road as far west as Notting Hill Gate. Much redevelopment of a similar nature, i.e. blocks of flats, but this time with shops underneath, spread along the Edgware Road, through Maida Vale and as far as Kilburn. Other districts where flat development took place where Hampstead, St John's Wood, Cricklewood, Balham, Streatham and, of course, Mayfair.

Housing developments, whether for sale or letting, seemed to follow, or in some cases lead to, the location of new underground railway stations. The era also marked a notable new kind of shop development. Previous to the 1914–18 War, there were practically no shopping parades developed outside of the established high street shopping positions, no matter in which town or city these were situated, however during the 1930s local parades of shops appeared.

The availability of coal, electricity and oil, the three sources of energy which powered the principal forms of transport, facilitated suburban growth, particularly that of London. From 1901, new railway stations were opened almost every year, commencing with West Ham and Goodmayes in the east of London, followed by the west and north-west. 1903 witnessed extensions of the lines to Chigwell, Harrow and Park Royal; in the following year to Uxbridge and Perivale; in 1905 to Letchworth and in succeeding years to almost every suburb within a radius of 7 miles; the last two stations, East Finchley and Barnet, were opened in 1940. Altogether the records of the London Passenger Transport Board reveal a total of 100 stations opened between 1901 and 1923 and 78 between that date and the beginning of hostilities. Construction went on even during the First World War, with some eleven stations being opened from 1914 to 1917. No further stations were opened until the extension to Welwyn Garden City, with the exception of the entirely new Acton Town—South Harrow line, which was constructed as an electric line. Until 1905 the trains ran on steam and the first line to change over to electric power was that from Baker Street to Harrow and Uxbridge. Another four lines were changed in that year and thereafter the pace of the changeover accelerated to almost 100% electrification by 1940.

Omnibuses and trams were in continuous use long before the turn of the century, the former sometimes horse-drawn and independently operated whatever the motive power. The principal operating company in London was the London General Omnibus Company Ltd (LGOC) which served the surburbs from the central main railway stations, Victoria, Waterloo, Kings Cross, Charing Cross and Liverpool Street. The City of London was served by buses starting from the Bank of England. The surburban termini included Putney, Cricklewood, Harlesden, Barnes, Ealing and Ilford together with intermediate stops.

By 1914 the bus system had been extended to cover no less than 70 different routes taking in Hampton Court, Reigate, Wanstead, Golders Green, Palmers Green, Southgate, Twickenham, Cockfosters, Epping Forest, Farnborough in Kent, Edmonton and the many districts between the suburban termini and Central London. In addition, there were many links between adjacent suburbs, which more than doubled the number of all bus routes.

3.8.2 The effect on Golders Green

It is impossible to trace the history of the development around every suburban railway station, tram or bus terminus although Jackson[19] has selected some dozen or so case studies on this particular aspect in varying degrees of depth. In particular, he has chosen Golders Green and some of the statistical information is indicative of what went on elsewhere.

The most prolific builder in the area was Edward Streather, together with his brother William and their sons, Reginald

Plate 5 **A shopping parade characteristic of the inter-war period, in Greenford, Middlesex. (Courtesy of Press-Tige Pictures Ltd.)**

Plate 6 This shopping parade in Hanwell, London, is also typical of the inter-war period. (Courtesy of Press-Tige Pictures Ltd.)

and Cecil. They built thousands of houses in Golders Green and Hendon, moving on to Edgware and Mill Hill in the 1920s and 1930s. Their houses were built in half-timbered style, tiled, gabled and 'cottagey' in appearance and these set the trend for the next 30 years of London suburban exteriors. Streather's three bedroom semi-detached houses in Finchley Road and St. John's Road were selling for £600 freehold, whilst Ernest Owers' houses were selling at £450 leasehold. On the Woodcock Estate at Montpelier Rise, Brady was charging £425 leasehold and at Golders Green F. Bostable was selling at the same price. In The Grove a slightly larger house was selling for £600 freehold.

Plot sizes were small, back gardens were short and rarely more than 80 feet deep. There was no overall plan, only compliance with the usual local authority requirements regarding road widths, drainage and the basic ruling byelaws with street layouts determined by the desire to pack in as many houses as possible. The pattern set was to be followed by other emerging suburbs over the next 30 years.

As the new houses were built, so the underground service was extended. The 12 minute interval between trains of 1907 became 10 minutes in the following year. A three minute peak hour service was introduced in 1909, and 114 trains (each way daily) at the beginning grew to 318 by the end of 1910.

In evidence before the Light Railway Commissioners who opposed a proposed tramway between Golders Green and Hendon in 1901, Golders Green was described as a 'place within three miles of Regents Park, where there are roses in the hedgerows and the larks are singing, a place almost unique in its rural character.' This 'unique place' was deliberately chosen for a new underground railway station. As soon as work began on the railway in 1902, agricultural land went up from between £150 and £200 to as much as £2000 an acre. Even before the station had opened, land near it was selling at the rate of £5500 an acre for shop development, and a few years later at £10000. Even half a mile from the station the Woodstock House Estate, with some 500 yards of frontage to the main Hendon Road, was sold in May 1908 at £1000 per acre.

Close to the station some nineteen houses were finished in 1905 and fourteen more in 1906. By 1907, the year the underground railway station was opened, some seventy-three houses had been built. Thereafter the number of houses finished were as follows:

1908	340
1909	461
1910	562
1911	744
1912	486
1913	514
1914	417
1915	432

By 1909 some two miles of new roads and six miles of sewers had been laid, and in the following year a further 2.25 miles of roads and five miles of sewers were completed. The number travelling to the new station in the first full year totalled some 1.5 million passengers, but by 1915 the total was over 10 million. At the beginning of the 1920s Golders Green was the fifth busiest station on the whole of the underground system with 35000 passengers travelling on an ordinary weekday — or about 1 million each month. By 1935 an additional 3 million brought the total number of passengers travelling to 13 million. In 1900 Golders Green consisted of a few large houses, some cottages and the farm itself, known as Golders Green. Although only three miles from Regents Park, those living there considered themselves to be in the country. In 1902 the Crematorium was opened and in 1906 the Underground Electric Railways Company of London Ltd acquired a site for a station. There had been proposals to open up the district around Golders Green with an electric tramway system, but only one line, from Cricklewood to Finchley, was allowed.

In 1893 a Private Bill had been approved by Parliament allowing for an underground railway from Charing Cross to Hampstead. The scheme failed to attract the necessary capital to carry it through, until in 1900 an American financier, Mr C. T. Yerkes, took over. He decided to extend the scheme and, in fact, made an application to Parliament for a separate Bill (this time not in the railway's name) to construct some 5.15 miles tapping Hendon and Finchley on the way to Golders Green in order to increase the traffic at Golders Green station. In addition, Yerkes' group obtained options on land adjoining the station site before the railway scheme was publicly announced. The line was eventually opened in June 1907, with two-car trains leaving Golders Green every 12 minutes from 5.17 a.m., to 12.31 a.m., taking 24 minutes to reach Charing Cross Station. Since building had not started in earnest at that time, the traffic at Golders Green was heaviest at weekends, with Londoners seeking not only the pleasures of the countryside, but the extra alcoholic drinking times which bona fide travellers living more than three miles from Golders Green were entitled to demand. As a result the service had to be extended to provide six-car trains every four minutes.

Here then was a case of a purely profit-motivated transport company promoting property development. Firstly, the company profited by dealing in land that was about to increase in value through the coming of the railway station and secondly it encouraged as much building as possible around the station, so as to capture rail traffic revenue.

3.8.3 Railways and builders

Yerkes' project was not, of course, the only example of a railway company's direct connection with developers on the level of finance. Between 1890 and 1908 Cameron Corbett contributed to the construction of stations in the Ilford area, at Hither Green and at Eltham Park. In 1912 Sir Audley Heeld, Sir Theodore Brinkham and Maple Trustees gave gratis part of the land for the underground's Edgware extension. Dollis Hill and Willesden Green together with Neasden were assisted financially by Messrs Andrews and the Dudding Park Estate Company. Subsidies from other developers assisted the development of the stations at Riddlesdown, West Weybridge, Runneymede, Petts Wood and West Wickham. This list grows longer and includes Hinchley Wood, Whitton, Woodmansterne, Stoneleigh, Belmont Halt and Hersham. At Woodmansterne the station building costs were £7000. The land was conveyed free and various offerings totalled £1500. Berty Fisher donated the land for Hinchley Wood, together with £2500, as well as purchasing £500 of Southern Railway stock.

The Southern Railway received £5000 of the total costs of £9900 required for the new station at Hersham between

Esher and Walton-on-Thames, Elm Park was assisted by Richard Costain and Sons. Albury Park and Falcon Wood were sponsored by New Ideal Homesteads. Records of all these transactions are contained in the *British Transport Historical Records.*

3.8.4 Finances of the underground and metropolitan railways

Lord Ashfield, the chairman of the underground railway company, in evidence before the Parliamentary committee on the Morden Extension Bill of 1923, contended that the full benefit of the initial capital investment outlaid on the expensive inner tunnel system could only be recouped by a policy of continued surface expansion outwards. In the 1936 evidence to the Barlow Commission, the London Passenger Transport Board gave figures for one mile without rolling stock. The tunnel sections cost £675000, the open section £275000, per mile. Speaking at the London School of Economics in 1934, Lord Ashfield revealed that his underground railways could pay only 1% on the *non prior* charge capital and that the average return on the whole capital was 3.25% with no provision for redemption. He advocated increasing the density of development, and pointed out that 12 houses to the acre were not enough to make his railways pay. This density provided only 24000 people within a half mile catchment area, averaging 3 million passengers per year per station. He calculated that at least another million passengers per station per year would be required to support a railway with stations a minimum of three-quarters of a mile apart.

Although all the railway companies dabbled in land speculation and obtained financial assistance from developers from time to time, only one railway company managed to adopt a positive land policy. This was the Metropolitan Railway which, under the Metropolitan Railways Act of 1885 and 1887, was able to form the Metropolitan Railway Surplus Lands Company. Under the 1874 Act the Metropolitan had enjoyed the neat distinction amongst other railway companies of being able to grant building leases and sell ground rents in respect of its lands in the City of London. The accounts showed a curious mixture of both land and railway income and in the endeavour to regularize the position, the company's ordinary stock was divided into ordinary revenue stock with the property hived off (a valuation being made) and the stockholders given £1 of new stock for every £2 of Metropolitan Railway consolidated stock held. In January 1919 yet another company was formed, to be called the Metropolitan Country Estates Ltd. This company purchased estates at Rickmansworth and Chalk Hill and some forty acres of railway land at Neasden were transferred from the parent company. This company, although it sold off some of its land to developers, went on to develop some 10 estates along the main line and the Uxbridge branch. It actually built some houses but in the main sold off the land, either to builders or in individual plots. Other estates purchased by the company were Kingsbury Garden Village, the Wembley Park Estate, The Cedars Estate at Rickmansworth and the Wells Estate at Amersham.

There is no doubt that these estate companies of the Metropolitan Railway had a powerful influence on the suburban growth. It is significant that the example of the Metropolitan Railway was not copied by the underground or the main line companies. Not only the Metropolitan Railway but all those associated with it, including building contractors, estate developers and financiers, formed a powerful combination and no doubt made it clear to possible competition that any attempt to obtain Government sanction to develop would meet with effective lobbying. In the event no other transport undertaking made any serious attempt to encroach on what was virtually the monopoly position enjoyed by the Metropolitan Railway.

There is no reason to believe that London and its suburbs presents a special case and doubtless there could be found evidence of a similar nature to explain the growth of other towns and cities in Great Britain.

3.9 SHOPS

An an interview in May 1978 Aubrey Orchard-Lisle, then the senior partner of Healey & Baker, commented on the growth of shopping parades in the new suburbs that had sprung up on the outskirts of London and Birmingham in the inter-war years. His firm acted for the specialists in parade development and were concerned in the purchase of road frontage close to the tube stations and forming part of the speculative housing developments. The financial success of these parades depended on securing multiple traders as tenants for the first lettings. The mutual attraction of multiple traders wanting to expand their branches and the developers securing readily saleable investments was one of the characteristics of the period between the wars.

3.9.1 'Multiple' shops

Although traders with more than one store or shop were in existence before the 1900s, the explosion of what later became known as the 'multiple' shop occurred during the 1920s and 1930s. The most important of all the multiple firms were the Co-Operative Societies. The estimated percentage share of total sales in the entire UK, between the years 1900—20 was:

Food and household requisites	7.5%	rising to 12.5%
Confectionary, stationery and tobaccos	1%	rising to 2.5%
Clothing and shoes	4%	rising to 6.5%
Other goods	6%	rising to 9%

Both Marks & Spencers and F. W. Woolworth, (the latter from America where it was known as a 'chain store group') had established themselves before the First World War. Other retailers were in the multiple shop category, such as J. Sainsbury the provision merchants, Boots the Chemist, Lilley & Skinner (shoes), and the Vestey Group (predominantly butchers) together with numerous smaller multiples where the chain of shops could be any number from five to thirty.

Two of the best known multiple chain stores, W. H. Smith, the newsagents and booksellers, and J. Sainsbury, have kindly provided information showing their expansion. In the case of Smith's they had by 1936 some 648 bookstalls on railway stations and 332 shops. Their retail turnover for the year 1925—26 was just over £4 million and this had increased to close on £4¼ million by 1936. Sainsbury's 159 branches in 1925 increased to 244 by 1938 but they were also involved in farming and food processing on a large scale. The history of their expansion is revealed in some detail in the following letters addressed to the author.

Archivist
Extension 18

W. H. Smith & Son Ltd
Station House
Harrow Road
Wembley
Middlesex HA9 6EP
Telephone: (01) 902 8831

2 November 1979
CH/57

Jack Rose Esq
28 Crawford Street
London W1H 1PL

Dear Mr Rose

DYNAMICS OF URBAN PROPERTY DEVELOPMENT: INDUSTRIAL REVOLUTION ONWARDS

In reply to your letter of 23 October, I enclose a sheet giving the information you requested. You will appreciate that every organisation has its peculiarities, and the figures should therefore be read with at least the folllowing in mind:

(a) W. H. Smith have a substantial wholesale trade; in the earlier years some of this was part of the trade of shops and even bookstalls.

(b) Bookstalls, most of which were on the railways or London underground, were/are held under licences of varying terms of between 7 and 21 years. When the London Transport contract was relinquished in 1973 the numbers were reduced overnight by 23 main and 63 substalls; similarly, when the British Rail contract was renewed in 1974 there was a drastic drop to 82 bookstalls plus associated platform stalls.

(c) I have presented figures for our main shops and bookstalls (main in the sense not of large but independently managed) believing this to show the truest comparison over the years; however, I have also set out the more complicated picture in 1965 so that you will have an insight into the more detailed situation. In general, the inter-war years saw an increase of substalls rather than main stalls, and the post-war years a realisation that rail travel was declining.

(d) In recent years there has been an emphasis on selling space rather than numbers of branches. Bradford Shop, opened in 1960 with 12,000 sq. ft, and Birmingham Shop, opened in 1973 with 29,000 sq. ft, cannot usefully be compared with pre-war shops on a one for one basis. A factor in the older shops is also the closure of the circulating library in 1961; this took up a comparatively large amount of space in many shops for very little financial return.

(e) The imposition of Purchase Tax in 1940, changed to VAT in 1973, boosted turnovers in later years.

Yours sincerely

Archivist

W. H. Smith — retail outlets and turnover

	1925–26	*1935–36*	*1945–46*	*1955–56*	*1965 25 Sept.*	*1975 31 Dec.*	*1979 3 Feb.*
Bookstalls (England and Wales) (numbers)							
Main only	615	648	578	515	259	65	62
Shops (England and Wales) (numbers)							
Main only	273	332	338	376	352	303	314
Bowes & Bowes Group				1	7	11	12
Craftsmith						2	8
Shops (overseas) (numbers)							
France	1	1	1	1	1	1	1
Belgium	1	1	1	1	1	1	1
Canada				4	10	36	55
Holland (50:50)						4	—
Retail Turnover (£)							
'000	4079	4221	8233	16838	30645	134036	258432

2 November 1979
T.W.B.J.

W. H. Smith — retail outlets at 25 September 1965

Bookstalls (England and Wales) (numbers)	
Railway	
main	243
sub.	236
sub. to shops	122
platform	86
Non-railway	
main	16
sub.	7
Shops (England and Wales) (numbers)	
Main shops (inc seasonal)	352
Sub. shops	14
Bowes & Bowes Group	7
Shakespeare Exhibition	1
Shops (overseas) (numbers)	
France	1
Belgium	1
Canada	10

2 November 1979
T.W.B.J.

J Sainsbury plc
Stamford House
Stamford Street
London
SE1 9LL
Tel: 01 921 6510

14 August 1984

Dear Mr Rose

Thank you for your letter in connection with your research project.

I have attempted to work out the number of retail outlets for the years you required - not an easy task as we only have a list of all our branches with all the known opening and closing dates. I think that these figures are accurate:

1925 159 branches
1935 218 branches
1938 244 branches
1945 234 branches
1955 257 branches
1965 230 branches
1969 242 branches
1975 216 outlets (including freezer centres and one petrol station)
1979 239 outlets (including 193 supermarkets, 20 independent freezer centres, 7 petrol stations)
1984 March 270 outlets (see below)

As for your second question, our gross receipts are classed as confidential information until 1973 when the company went public. I enclose a General Information Sheet which details Before Tax and After Tax Profits from 1973.

Yours sincerely

Bridget Williams
Company Archivist

Sainsbury's General Information
Sainsbury's is a London-based multiple food chain established in 1869. Branches are mainly concentrated in the southern part of the country, the furthest from London being Exeter in the West; Southport in the North West; Leeds and Bradford in the North; King's Lynn and Great Yarmouth in East Anglia. Sainsbury's now have 270 outlets, of which 242 are Supermarkets, 12 are independent Freezer Centre and 13 Petrol Stations. Sainsbury's became a public company in July 1973 and employ about 58000 staff, of which 33700 are part-time.

Turnover and profit

	Turnover (£000)	*Profit*	
		Before tax (£000)	*After tax (£000)*
1973	296900	11400	6900
1974	362100	13600	6300
1975	452800	14600	6900
1976	543400	15400	7400
1977	663800	26200	20200
1978	811100	27600	21200
1979	1 007000	32700	26400
1980	1 226600	46000	35100
1981	1 589000	65800	49900
1982	1 950500	89100	68000
1983	2 315800	107300	73300
1984	2 688500	138100	89000

Supermarket statistics
The average sales area of Sainsbury's supermarkets at the 1983/4 financial year end (24 March 1984) was 16070 sq. ft, of which approximately 90% is given over to food (the average size of the Stores operated by eight of the principal multiple groups within the UK, including Sainsbury's, is 7000 sq. ft). Fifteen Sainsbury's supermarkets (average sales area 25 530 sq. ft) were opened in the financial year 1983/4.

The selling area in Sainsbury's supermarkets has increased as follows (figures given in sq. ft):

1973	1337000	1979	2688000
1974	1711000	1980	2766000
1975	1936000	1981	2978000
1976	2240000	1982	3282000
1977	2391000	1983	3564000
1978	2538000	1984	3944000

The average Sainsbury supermarket turnover is about £214200 per week compared with a weekly average turnover of all stores operated by the principal multiple groups of approximately £22500. The average turnover per square foot of selling area in Sainsbury's supermarkets is about £13.58 per week, compared with approximately £6.05 for all multiples. The average number of checkouts per Sainsbury's supermarket is 17. The total number of lines stocked is approximately 9000. Most of the supermarkets now have wines and spirits.

Own label trading

Sainsbury's 'own label' business now accounts for well over half the grocery trade, although the sale and promotion of manufacturers' brands, offered side by side with Sainsbury's brands, remain of vital importance. Sainsbury's offer own label lines in their major stores. This figure is calculated on a count of all packet grocery and canned good lines and includes non-foods, wines and spirits, and some pre-packed perishables such as margarine, yogurt, cheese, bacon, sausages and pies. In addition, there are 500 freezer centre lines and 770 items in the range of family clothing.

Quality control is an important aspect of Sainsbury's own label business and is the joint responsibility of the buyers and of the laboratory chemists and bacteriologists. Upwards of 60000 routine analyses are made in the course of a year in Sainsbury's London laboratory.

Freezer centres

Sainsbury's opened three new freezer centres in 1983/84,

making a total of 23. While it remains policy to open 'independent' freezer centres most of the expansion in this area is within Sainsbury's supermarkets.

Range extension

Sainsbury's successful range extension has been in large measure due to the company's willingness to respond to the changing needs of customers. In 1950 the average number of lines was 500 rising to 4000 in 1965. Today, Sainsbury's largest stores carry up to 10 000 different products, the vast majority of which are foods. Particularly notable are the in-store bakeries and delicatessen counters of larger branches, the free-flow fruit and vegetable displays which allow the customer to select the size and quantity of produce, and wine departments stocking up to 200 different wines. Sainsbury's is the biggest wine merchant, butcher and greengrocer in Britain today.

Central control

A high degree of centralized control has for many years been the policy of the company. Virtually all goods are bought centrally; retail prices, special promotions and displays are fixed and organized centrally. Sainsbury's introduced a computerized stock replenishment system in 1961 and were the first retail food company to do so in England. In 1971 a new ordering system for stock replenishment using data capture equipment was developed and this system, which offers very considerable advantages in speed, accuracy and cost saving, is now in use in all outlets.

Distribution

Until the mid-1960s, daily deliveries to all branches were made from a central depot in London. With the growth and spread of the business, the company made a deliberate policy to move the base of its operation from Central London to avoid traffic congestion in the centre of the metropolis and thus four main Sainsbury depots were established around London, which has proved environmentally beneficial. These are at Basingstoke in Hampshire, Buntingford and Hoddesdon in Hertfordshire and Charlton, a suburb of South East London. The total floor area of these four depots is nearly 1½ million square feet (34 acres).

A great deal of prepacking of perishable items such as cheese, bacon and fresh meat is undertaken at the depots which, in effect, operate as an extension of the branch preparation areas. Since the initial investment to build and equip these depots there has been a continuous programme of investment to extend and update the facilities. In addition to the Sainsbury depots, a number of contractor depots supplement the company's distribution system, which ensures full coverage of the area where shops are located with the minimum journey time for vehicles.

Subsidiary and associated companies

SavaCentre Ltd SavaCentre, which is jointly owned with British Home Stores, opened its first hypermarket in November 1977, in Washington, Tyne and Wear. By March 1984 there were five SavaCentres, which an average sales area of about 75000 sq. ft. A profit of £8.8 million was declared for the financial year 1983/4.

Homebase Ltd Homebase is a subsidiary of Sainsbury's, launched in 1979. The company is owned by Sainsbury's (75%) and by the Belgian company GB-INNO-BM (25%). By the end of the 1983/4 financial year there were 14 Homebase stores, with a further 9 projected to open during the year 1984/5. The average sales area of these stores is about 45000 sq. ft.

Haverhill Meat Products Ltd The company is jointly owned by J. Sainsbury plc and Canada Packers Ltd of Toronto. Its factory at Little Wratting near Haverhill in Suffolk has produced Sainsbury's Tendersweet Bacon and Meat products for over 10 years. Under the terms of an agreement signed in 1969 manufacturing operations have widened to produce fresh pork, fresh and cooked sausages and meat pies.

J. Sainsbury (Farms) Ltd Sainsbury's operate two farms in Scotland. One is centred on Kinermony (a total of 700 acres) where the Kinermony herd of pedigree Aberdeen Angus cattle was founded in 1945. The other is a commercial farm of 1000 acres in Inverquhomery, which is half arable and half fattening cattle.

Breckland Farms Ltd In January 1976 Sainsbury's and Pauls and Whites Ltd formed a new company (Breckland Farms Ltd) to undertake large scale commercial production of quality pigs in East Anglia. The project was intended to augment the supply of pigs to Sainsbury's to meet the additional demand for pork products from new supermarkets. The development extends a link between the companies built up over a number of years which culminated in Pauls and Whites becoming one of the principal suppliers of pigs to Sainsbury's.

J. Sainsbury (Properties) Ltd This subsidiary manages, but does not own, the company's properties.

Kings Reach Investments Ltd The Kings Reach development is situated on the south side of the River Thames close by Sainsbury's Head Office in Stamford Street. In addition to Sainsbury's the shareholders in this company are the Union International Company plc, the Stock Conversion and Investment Trust plc and a subsidiary of Reed International plc.

Shaw's Supermarkets Inc. (USA) In November 1984 the company acquired a 21% minority shareholding in Shaw's Supermarkets Inc., a very successful private American food retailing company with stores in Massachusetts, New Hampshire and Maine.

From the exhaustive supply of statistics a picture of substantial growth emerges, both in the number of branches opened by multiples and the consequent capture of a considerable volume of the total retail sales effected. Tables 3.4 and 3.5 tell the tale quite explicitly. Of all retail sales in the main commodity groups the multiples' share increased from an average of approximately 4% in 1900 to some 19% by 1939. The trend towards the growth of multiple shops

Table 3.4

	Firms with 10 or more branches		*Firms with 25 or more branches*		*Total number of all multiple shop branches*
	Firms	*Branches*	*Firms*	*Branches*	
1900	257	11645	94	9236	11645
1920	471	24713	180	20602	24713
1939					44487

resulted in keen competition by them for premises in the established high streets throughout the United Kingdom. This resulted in a gradual but nevertheless drastic alteration in the composition of high streets almost everywhere. The nature of shopkeeping before the advent of the multiples to any great degree was for the shopkeeper to own his own property, generally freehold, and to use the premises above the shop as his residence. Certainly during the 1930s a sustained effort by various estate agents acting on behalf of the multiple groups in competition with one another descended on high streets of towns and even villages, tempting property owners, especially where they were owner-occupiers, to sell to their clients. Specialist firms, almost all of whom were situated in the streets around Hanover Square in London, became well known for the success of their activities in this direction. Amongst the better known were Hillier, Parker, May and Rowden of Grosvenor Street, Healey & Baker of St George's Street, Hanover Square, and Dudley Samuel and Harrison of Bruton Street.

Table 3.5

	Increases in branch opening for firms with	
	10 or more branches (%)	*25 or more branches (%)*
1921–25	20	19
1926–30	21	25
1931–35	12	13
1936–39	11	13

3.9.2 Departmental stores

The last quarter of the nineteenth century had witnessed the advent of the departmental store which owed its origin to the changes in women's fashion. At around the turn of the century, William Whitely in Bayswater and Gordon Selfridge in Oxford Street pioneered the new concept by the construction of a large purpose-built building which not only provided for retailing at ground floor level but also on many floors above. The provision of lifts and wide staircases, lofty space and open-plan floors enabled this kind of store to stock a great variety of merchandise and presaged the very large number of departmental stores that subsequently were built in London, Birmingham, Manchester. In fact, in most of the principal towns in Great Britain almost any departmental store of size became the focal point for shopping in the high street in which it was situated.

3.9.3 Warehouses

Whether it was a departmental store or a specialist shop selling goods that were not manufactured on the premises, these goods had to be purchased by the shopkeeper and the normal procedure until well after the Second World War was to obtain these goods from warehouses. The distribution of goods of all descriptions was necessarily a two-tier process. Large multi-storeyed buildings had existed for some considerable time in the centres of London, Manchester, Birmingham and in other important locations. The warehousemen generally specialized in stocking fairly large quantities of goods of a particular type. Thus there were warehouses which dealt only with cotton goods, some with kitchen and allied equipment, others with furniture, etc. In some cases, one of the floors, usually the ground, was used as a showroom where one of each article was displayed. The remainder of the building was used for storage and despatch. Shopkeepers made special journeys, sometimes from long distances, to choose stock for their shops. Some warehousemen employed travellers to call upon shopkeepers at their own premises, either with catalogues or with samples of goods held at the warehouse. The owners of the warehouses would, of course, place large orders with manufacturers and maintain a source of continued business. There was a tacit understanding that manufacturers would not sell directly to shopkeepers for fear of offending their better customers, the warehousemen, who would thereby lose business and possibly switch to other manufacturers in retaliation. The warehouses which were in existence in early Victorian times rarely had lifts. They dealt with the problem of importation of the goods in bulk into their premises and their subsequent distribution by having doors cut in the walls of each storey, with a gantry or crane above.

These wholesale warehouses needed to be in easy reach of such transport facilities that existed, notably railway stations, and since the railway stations themselves were generally in the centre of cities, this is where these warehouses were to be found. An adjunct to the distribution problem of the times was provided by the numerous transport firms who were employed both by the warehousemen and the shop owners to transport the goods from the warehouse to the various shops. The transporting of goods from the manufacturer to the warehouse was part of the manufacturer's service but the advent of the multiple chain store rapidly changed this system, since the proprietors of a chain store with sufficient shops could match the order of any warehouse. Consequently the provision of the warehouse owned by a chain store proprietor, and now called a distribution centre, came into existence. The invention of the pallet and fork lift truck together with improved road transport produced the distribution centre of single-storey construction. Better roads gave freedom of choice of location of these distribution centres away from the traffic-intensive city centres. Roads were used to such an extent that motorway systems were developed to cope with the volume of goods traffic captured from the railways.

Many of the old warehouses, particularly those situated in the City of London and its fringes, were used as office buildings, particularly before the 1947 Town and Country Planning Act. Since 1 July 1948, when the Act became operative, planning difficulties caused a standstill in this operation. However, many of the better positioned warehouses were acquired by property development companies who revamped them, adding lifts, central heating, floor coverings, replastering where previously there was only fair-faced brickwork and generally giving them the appearance of a conventional office building. With the scarcity of office accommodation that existed after the Second World War it was not uncommon for some tenants to lease and occupy these buildings, using them purely for office purposes. Until the 1947 Act was amended in 1968 it was lawful to continue the use of these premises as offices if the use had commenced before 1 January 1964 and had extended to a term of four years without the planning authority giving notice to the tenant to cease the office use.

REFERENCES

1. *Journal of the Town Planning Institute* No. 5, 60 May 1974.
2. Cherry G. E., (1974) *The Evolution of British Town Planning*, Leonard Hill, Glasgow.
3. LCC (1939) *London Statistics*, 41.
4. LCC (1939—46) *Statistical Abstract for London*.
5. LCC (1944) *Greater London Plan*, Appendix 2.
6. Interim report of The Departmental Committee on Regional Development 1931 (Cmmd 3915).
7. Read, D., (1972), *Edwardian England*, Historical Association, London.
8. *The Journal of The Town Planning Institute*, XVIII, 1931—1932.
9. Cullingworth J. B., (1975) Reconstruction and Land Use Planning, 1939/1947. In *Environmental Planning 1939—1969*, Vol. 1, H.M.S.O.
10. Cherry G. E., (1974) *The Evolution of British Town Planning*, Leonard Hill, London p. 121.
11. Cullingworth J. B., (1975) *Environmental Planning 1939—1969*, Vol. 1, p. 259.
12. Hunter Committee report (Cmmd 9235).
13. Salisbury Committee report (Cmmd 658).
14. Rent and Mortgage Restrictions Act 1923, which followed the Onslow Committee Report (Cmmd 1805).
15. Increase of Rent and Mortgage (Restrictions) Act 1919.
16. Increase of Rent and Mortgage (Restrictions) Act 1920.
17. Rent and Mortgage Restrictions (Continuance) Act 1923.
18. Ridley Committee report (Cmmd 6621).
19. Jackson, A. A. (1973) *Semi-detached London*, Allen and Unwin, London.

FURTHER READING

Aldridge, T. (1980) *Rent Control and Leasehold Enfranchisement*. 8th edn. Oyez, London.

Buchanan, C. D. (1963) *Traffic in Towns*. HMSO, London.

Department of Environment (1973) *Passenger Transport in Great Britain 1973*. HMSO, London.

Heap, D. (1973) *An Outline of Planning Law*. 6th edn. Sweet and Maxwell, London.

Krimgoltz, M. and Good, B. J. C. (1982) *Renting and Letting a Home*. George Goodwin, London.

Lave, R., Powell, T. J. and Smith, P. P. (1971) *Analytical Transport Planning*. Gerald Duckworth and Co., London.

Leibbrand, K. (1970) *Transportation and Town Planning*. Leonard Hill, London.

Lichfield and Proudlove (1976) *Conservation and Traffic: A Case Study of York*. Sessions Book Trust, The Ebor Press, York.

Megarry, R. E. (1970) *The Rent Acts*. Stevens and Sons, London.

Mumford, L. (1964) *The Highway and the City*. Secker and Warburg, London.

Pettit, P. H. (1981) *Private Sector Tenancies*. Butterworths, London.

Partington, M. *Landlord and Tenant*. Weidenfeld and Nicolson.

Pressat, R. (1973) *Population*. Penguin, Harmondsworth.

Ratcliffe, J. (1974) *Introduction to Town and Country Planning*. Hutchinson, London.

Telling, A. E. (1977) *Planning Law and Procedure*. Butterworths, London.

Young, A. P. (1970) *Transportation for Towns of the Future — Urban Renewal 1970*. University of Salford, Salford.

4

The Economics of Property Transactions after 1945

4.1 INTRODUCTION

Although it may be convenient to regard the post-war era in terms of four distinct periods the dates by which they are identified do not act as cut-off points from either the legislation or economic factors of the previous or subsequent periods. The era to date has witnessed the intensification of the differences both in the philosophical approach to planning and the attendant taxation on the profits of development. Paradoxically, due to the effects of inflation, property owners have benefited more during Labour than Conservative administrations. Draconian measures enacted by Labour have never remained on the Statute Books long enough to bite into profits before a switch in government brought about a timely reprieve. The financial sections of the 1947 Town and Country Planning Act and the Land Commission and the Community Land Act disappeared, although the 1967 Leasehold Reform Act and several of the Housing and Rent Acts remain operative with only marginal alterations. Yet it was a Conservative Government that initiated the Capital Gains Tax legislation, was responsible for the freeze on the rents of business property-which precipitated (but was not the cause of) the banking crisis, and legislated for punitive rates on empty commercial buildings for motives of political expediency, bordering on panic.

The post-war era has witnessed a phenomenal expansion in the field of communication and the proliferation of commentators specializing in diffusing information concerning property. Both national and local newspapers devote substantial space to all aspects of property development, since planning applications are open to public inspection. Estate agents seeking publicity voluntarily supply details of transactions and contribute articles on trends of supply and demand, and place advertisements which in some cases splash across several pages. Stockbrokers specializing in the shares of quoted property companies employ analysts who not only circularize their clients with detailed appraisals but also contribute to financial journals which regularly include a property section. There are also numerous magazines (delivered free with the morning newspapers) which carry property advertisements and the inevitable commentaries.

Some 60 000 or so qualified surveyors and valuers are members of professional institutions which publish journals. Additionally, there are specialist journals concerned variously with valuation, town planning, property taxation, property management and the like and even such productions entitled *What Mortgage*, *What Investment*, etc., sell 30 000 copies monthly. The most prestigious of the numerous publications is the *Estates Gazette*. Pre-war copies extended to a dozen flimsy pages, the modern glossy version is three-eighths of an inch thick. All this publicity has had inevitable reaction, in a

Plate 7 Bomb ravaged Cannon Street, with St Pauls in the background. (Courtesy of Popperphoto.)

Plate 8 Brittanic House, Moor Lane, London — BP's Headquarters, a 1960's building ahead of its time. (Courtesy of Popperphoto.)

society where one half at least, is restricted financially to a weekly pay packet.

There is understandable curiosity, doubtless tinged with envy, about those who benefit from the profits generated from the multitude of millions entailed in these transactions. Notwithstanding the existence of a Labour Government in the immediate post-war era, the mood of the public was to encourage the resurgence of a prosperous Britain and if industrialists, exporters and manufacturers made handsome profits, but their incomes were taxed at the highest possible level, their success was considered socially acceptable.

As will be shown on page 151 the one sector that made substantial profits, yet quite legally avoided tax, created a relatively small group of individuals, each of whom became a millionaire, in the space of a few years. Oliver Marriott's *The Property Boom*[1] lists their names and even some details of their transactions. The majority of these men were content to maintain a low public profile but it needed only a few extroverts to seek personal publicity. This is not to suggest that there did not exist those who were undoubtedly gifted with a flair for recognizing the potential of selected property as purchases for investment or development, and accepted by the financial institutions as excellent borrowing risks. Understandably as so often occurs, the bahaviour of the few publicized by the media highlighted the seemingly unjust situation and the entire property industry came under political scrutiny and attack.

In examining the main factors that distinguish the four periods since the Second World War the single thread that runs through them all is the public reaction to development and the people who profit from it. In their defence, it is patent that the volume of new commercial buildings, if not houses, would never have been built except by speculative private enterprise. They gave work to the many building operatives and in many cases replaced older properties as well as rebuilt bomb-sites. For those who worked in offices and factories they provided healthier and more pleasant conditions. If the prevailing tax gave the property developers rewards inconsistent with their efforts, it seems the more rational remedy would have been to have taken the long-term view that is envisaged in the provisions of the Capital Transfer Tax legislation: whatever an individual accumulates comes back to the State in no small measure when he dies. There are more apparent factors that affect the entire era, namely the phasing out of the residential investment sector that was in private hands; the oversupply of commercial property resulting from the cornucopia of the money supply and the see-saw effect of legislation and its repeal that has so far prevented the property industry and the public it ought to satisfy from reaching a rational balance.

The first of these four periods extends from the end of the war until 1954. Building restrictions caused properties to mirror the state of the motor car market for some years after the war when new cars were almost impossible to obtain and secondhand cars dating from before the war were selling at higher prices than the new models; thus the market for property generally was in secondhand goods.

Of greater importance socially was the revolution that affected the occupiers of low rented houses held as investments by large scale property owners. The London Auction Mart witnessed in the years immediately following the election, sale rooms occupied every day, five days a week, by auctioneers selling off whole streets of houses and housing estates to buyers who subsequently fragmented them by individual sales to tenants much in the same way as a shopkeeper would buy goods at a wholesale price, making a profit on the inflated prices he would charge individual customers. The Labour manifesto, published before the election, had made it clear to the builders and developers of housing estates that there could be no profit in building houses to let. Equally, the retaining of rent control with perhaps minimal relaxation made it prudent for large scale investors in housing to sell and re-invest elsewhere.

This secondhand market was, therefore, a seller's market. Men and women released from the armed forces and war work had no difficulty in finding employment in the factories, shops and offices where labour had been in short supply for so long. The accumulated wages and pay earned (which during the war had few outlets) and the gradual relaxation of wartime restrictions on the production of items for domestic consumption, gave a boost to trade and accelerated the demand for all types of property. The government was intent on retaining strict control on commercial property development until it had enacted its programme. It would seem that it was content with the situation that housing for rent supplied by the private sector would be minimal if not non-existent and it had plans for dealing with this demand with municipal housing.

The beginning of the second period could be pinpointed to that date in November 1954 when at the Lord Mayor's Banquet at the Guild Hall, Mr Churchill announced that all building restrictions were to be at an end, thus launching the building boom which was to last until the end of 1964 when George Brown announced his 'ban'. The third period could be described as a financial boom which although attended by some hiccups nevertheless persisted until 1973 when it ended with the crash that threatened the banking establishment. The fourth period is one in which there has been a reversal of the situation in the immediate post-war years in that commercial property is in greater supply than seems warranted by the demand, although as if to prove the rule there are exceptions due to tax concessions to encourage the constriction of industial buildings and the relocation of high technology operations and specialized office users away from traditional centres.

There are, of course, overriding factors that affected the entire era, principally taxation closely aligned to planning legislation, recessions in world trade, trends in transport and growing pressure from the public for a voice in planning decisions, the direct investment of massive institutional funds into property and, overshadowing these, the philosophies of the main political parties.

4.2 THE SECONDHAND MARKET, 1945–54

With the end of the war a Labour government was elected which was committed to the establishment of a welfare state and the divestment of most of its colonial possessions. It also proposed legislation which was intended to go a long way towards redistributing the wealth of the rich including, or course, the property owners. The embargo on any new building, whether for housing or commercial use, except in special circumstances remained in full force, controlled by a licensing system operated by the Ministry of Works. It was considered essential that the economy of the country be resuscitated by any means but particularly by manufacturing for export. At the same time war rationing continued on a large range of items including food, so much so that after the war had ended bread was rationed for the first time; petrol was available only on a coupon system; and electricity and gas were rationed by cutting off supplies from time to time. This created what can only be described as a siege economy.

Whilst there was a free market in the sale and purchase of houses it was illegal to demand a premium for rented accommodation below a certain level of rateable value and this type was in the main regulated to the rent paid during the war. There were no restrictions on transactions of commercial properties but to protect the stock of residential properties in certain areas, notably in the centre of London, it was made illegal to use any premises that had previously been used as living accommodation for business purposes. The Defence Regulation 68 CA which spelt out the rules and penalties became a talking point for those agents concerned in areas where offices, in particular, were in demand.

The market in residential properties was on three tiers. So far as occupiers were concerned the lower bracket new housing consisted of prefabricated bungalows erected by Government agencies and let at low rents through the local authorities. Many empty houses vacated by those who had left for reasons of safety during the war were requisitioned, the local authority paying rent to the owners and then subletting. Those who held houses as an investment were faced with low controlled rents but increased liability for maintenance and while it was forbidden to spend money on repairs it was considered inevitable that rent control would continue after building restrictions were removed.

Since there was no control on selling prices as houses became vacant they were sold rather than re-let. The permitted amount an owner or tenant could spend on building repairing or even decorating a property was at one period as low as £10. This limit was gradually raised as conditions improved, until November 1945 when it became no longer necessary to obtain the licence hitherto necessary (which incidentally specified the spendable amount). Other controls applied to landlords extended to tenants who were prohibited from charging a premium or key money when transferring a tenancy and there is no doubt that a considerable black market operated as a result.

Most returning servicemen and women were forced to accept a lower standard of accommodation than was available pre-war and many had to be content with furnished houses and flats pending a return to normal conditions including the promised municipal developments. The alternative was to purchase houses at prices which were approximately double the pre-war prices during the first few years after 1945, and which later increased with inflation.

During the war the many refugees from Germany and other countries were forced, for security reasons, to reside in London, and congregated in particular districts. Since the advent of the seventeenth century, when French protestants settled in Spitalfields, immigants have commonly formed tight knit communities: the Russian and Polish Jews in Whitechapel, the Italians in Clerkenwell and the Cypriots in Soho were pre-war examples. The war had forced changes and it was the custom of at least one bus conductor to announce at a particular stop, 'Swiss Cottage — have your passports ready please.' It is fair to comment that immigrants from European countries have integrated into the community and immediate post-war conditions gave no indication of the problems that were later to occur due to the influx of non-Europeans, who for reasons of education, religion, colour and culture have caused problems in London and other towns.

Until 1939 almost all speculative building had occurred in the residential sector. With the exceptions of a few trading estates, factories were built for manufacturers who bought the land and employed contractors to build. Office blocks were generally owner-occupied although there were some developed for letting by a few investment companies, mainly in the City of London. The same conditions applied to warehouses and the building industry was predominantly engaged in producing houses and flats. The latter part of the nineteenth century had witnessed the development of the inner, and the inter-war period the outer, suburbs. Originally the developers had planned to sell the finished houses to owner-occupiers but with the surplus they created and with the demand for an outlet for investment funds, whole estates were sold to insurance companies and other funds who bought them after they had been let.

The tenants rarely entered into leases, more usually they paid rent weekly and left it to the owners to discharge rates and be responsible for repairs and insurance. This state of affairs was the exact opposite to post-war commercial lettings, when it was common for landlords to obtain for each building one tenant who took a long lease and who would undertake to pay all outgoings. Thus the owner-investor could enjoy a virtually trouble-free investment, considered by some preferable to investing in Stock Exchange securities showing a comparable yield and Government stocks at a lower yield.

This preference sparked off a huge disposal of residential investment property creating a new property market dealing in 'weeklies'. The London Auction Mart, when situated in Queen Victoria Street, had the appearance best described as resembling a school with large classrooms on each floor. Auctioneers could rent a room by the day and conduct sales. The volume of business was such that some seven rooms were in constant use five days a week and saw the transfer of weeklies in lots of several hundreds down to just two or three houses as small investors joined the larger vendors in the rush to sell. The buyers were dealers regarding themselves as wholesalers who after purchasing them proceeded to operate a 'buy out' or 'sell out' campaign. Each of the tenants would be approached and offered the alternative of purchasing or vacating his own house, taking with him a substantial sum enabling the operator to sell in the open market. Normally the operator would choose to buy out at least one of the tenants as a first move. The price he then obtained from the open market sale would, of course, be higher than that of a tenant purchase. However the difference between the two prices was an attractive inducement for a tenant to buy, since he was guaranteed a profit if he resold and since plentiful mortgage facilities were available a significant change in property ownership occurred.

Similar activity occurred in the disposal by large investors of blocks of flats (including those buildings known as mansion blocks), some built before the turn of the century. At this period the break-up into long leasehold tenure of individual flats was not considered, mainly because building societies and other mortgagors were averse to lending on this type of security. However a not inconsiderable business occurred in the acquisition of blocks, buying out tenants, and converting the larger flats into two or more smaller units. The entrance halls and common parts were redecorated and the converted flats equipped with new bathrooms and kitchens. The incomes of the blocks were increased and formed the basis of mortgage arrangements, often equalling the capital involved and producing a satisfactory equity for the developer. Unlike the idea of lending on a leasehold individual flat, there was evident willingness to lend on an entire income-producing block.

The activity in commercial property could be said to have depended on the inability of the insurance companies and other funds to realize the power of money. In essence, they were content to advance money on mortgage to speculators in property at low rates of interest for long periods before redemption. The sums involved ran into many millions and any over-valuation at the start was corrected by inflation long before redemption date. Institution investment, other than by mortgage, first found its outlet in shop property mainly in single units in important shopping streets and where the shop was let to a multiple branched tenant adjoining others of comparable covenant value.

The concept of inflation was almost entirely ignored so most leases granted were for 21 years and, where the tenant could be persuaded, 42 or even 99 year terms were fixed without rent review. At a time when the Chancellor of the Exchequer, Dalton, was able to dispose of Government stock at 2½% a yield of 5% on shop property was considered satisfactory. In the City of London, the West End of London and the centres of towns of the calibre of Birmingham, Manchester, Cardiff and Glasgow, considerable transactions were seen in vacant office and other commercial buildings of all sizes. Property dealers snapped them up, let them mostly on single tenancies and resold to institutional funds who would invest at around 6%, depending on the quality of the tenant, thus regarding this type of investment somewhat inferior to first class shops. In many cases the high tax bracket in which property dealers found themselves led to the alternative system to selling which subsequently resulted in the emergence of a proliferation of publicly quoted property investment companies.

Given the opportunity to purchase a vacant office building, an operator would estimate the rent he could achieve from a single tenant and in the then state of the market would probably have succeeded in agreeing a price to show a 10% return. Few, if any, institutional funds would have considered this type of transaction appealing and his only competitors would have been an occupier or another dealer. Once let, the property would be valued on a 6% basis by a mortgagor who would then lend two-thirds of this value, enough to cover the original capital outlay. If the property was purchased at £100 000, let at £10000 per annum and subsequently revalued at £166 700 the mortgage would have been £111 138. Interest on this sum at 5% is £5557 per annum, leaving £4443 per annum in the hands of the operators who, having retained and not sold the property paid no tax on the uplift in value although the equity income of £4443 per annum was subject to income tax.

One variant in the methods to avoid tax was for each property to be purchased in a separate company, the entire shareholding being owned by the dealer. Instead of selling the property after its financial development, tax laws operating at the time enabled the sale of the shares to occur without liability to tax, provided the operator did not repeat the process often enough to be judged a dealer in shares.

Another variant of the financial structuring of property dealing had as its basis the requirements of the funds to invest in securities comparable to Government stocks, considered as the ultimate in safety. Chief amongst the alternatives available were ground rents and the funds were willing to buy at 1—1½% over the 2½% Government stock. It became profitable for operators to buy a freehold and actually create a leasehold interest of 99 or in some cases 999 years selling off a freehold ground rent of up to 33⅓% of the projected income and retaining the leasehold interest as an investment. In many cases the same fund would buy the freehold ground rent and give a mortgage on the leasehold, the total effect of which was increased profitability to the operator.

Until 1948 when the 1947 Town and Country Planning Act came into force, warehouse premises in suitable locations and purchased at lower prices than established office properties suffered no restrictions if the use was changed to office occupation, thus opening up another opportunity for exploitation.

Mention has been made of the necessity to create investments favoured as suitable securities for the funds who generally insisted on purchasing or mortgaging buildings let on a single tenancy basis, with the tenant entering into a full repairing and insuring lease, paying a rent exclusive of rates and all conceivable outgoings. Most office buildings were in multiple occupation with the tenants on short leases often on a monthly or quarterly basis. Except for shops, where the tenants had limited protection under the 1927 Act, business tenants had neither security of tenure nor the benefit of rent control and this made it easier for operators to obtain their objectives, until the Landlord and Tenant Act of 1954 was enacted. New buildings were restricted to manufacturers and exporters, including other firms who could persuade the Board of Trade to support an application to the Ministry of Works for a licence, not only to build but within stipulated price limits.

There were, however, several new and large office buildings that came to be erected under what was described as 'lessor schemes'. The Government was in urgent need of office accommodation and made it known that any developer with a suitable site would be given a building licence subject to an occupational lease for a Government department. The average rent payable was eight shillings per usable square foot of the completed building. With the abundant supply of bombed sites and building costs below £3.50 per square foot of usable space, the developer could obtain around a 10% yield on his total outlay if the price of the land was not unreasonable. In the event, most of these developments occurred in secondary office locations where land costs allowed for this measure of yield. With the end of the bombing, some months before the war was declared over, prospective developers were thinking of eventual redevelopment of commercially located sites and the repair of unusable buildings. The Government had instituted the War Damage Commission, charged with administering payments due to property owners either to repair damaged buildings, the sum being the necessary cost of the works, or to agree to a 'value payment' generally where the building was razed to the ground or incapable of mere repairs. Owners who accepted value payments were more often than not anxious to sell the property. Others who were eligible for 'cost of works' compensation were frustrated by building controls. Thus there grew up another speciality market for these types of property. Astute developers were able to take advantage of the one concession granted to them by the Ministry of Works. Where the district surveyor could be persuaded that a building left in its existing state was dangerous to the public at large because it was likely to collapse, he would issue a 'Dangerous Structure Notice' the effect of which was to compel the owner to repair it. Armed with such a notice the Ministry automatically permitted its repair. Since the issue of these notices by district surveyors was entirely at their discretion, one or two at least were not adverse to the inevitable temptations to exercise their discretionary powers.

In summary, the period might well be described as one where the relief felt at the return to peace was tempered by the resignation that for the present all had to be satisfied with making the best of what was available. Property transactions although profitable for some, left most people not already settled in houses and businesses in limbo, having to remain content for the time being with promises by the Government for better times to come.

What was clearly evident, however, was that residential rented accommodation was not to be supplied by the private sector, that massive amounts of money were to be made available for non-residential development and investment and that the institutions had not shown any intention of applying the power their money gave them by developing or investing directly into property, but preferred to finance by lending to property companies and individuals.

4.3 THE BUILDING BOOM, 1954–64

The decade that commenced with the abolition of building licences marked the end of most if not all other restraints on commercial activity. With a Conservative Government in power and the accent on the production of goods of every description for domestic consumption as well as for export there was for the first time since before 1939 a return to a *laissez-faire* economy.

The spring of 1954 witnessed an increasing number of

licences issued by the Ministry of Works and several building projects had commenced. Alerted to this trend, entrepreneurs (not necessarily property owners) were considering the purchase or the taking of building leases of bombed sites in the centres of the larger cities and towns. London, of course, was desperately short of office accommodation and showroom and workroom space. Many provincial towns had lost the shops that were previously in their centres and were envisaging comprehensive redevelopment of completely new streets. The all-clear signal had been sounded for an unparalleled spate of building and there was room for both established development companies or like-minded individuals to participate in the bonanza.

Wholehearted encouragement both at national and local Government level to ensure that the developers could get on with the job permeated to the planning authorities, who gave near immediate consent to plans — in some cases the official stamp could be obtained within a week or so. Cases are on record of architects calling on the planning officer, obtaining his recommendation thereafter approved by the planning committee and the full council meeting all within the space of a few days. If there were conservation groups in existence their voices were either not heard or if they were, not heeded.

The personal recollections of the author are typical of the experiences of many in this era. An estate agent instructed to sell a site would telephone a prospective purchaser who, with no further details than the address, price required and the approximate size, would take a taxi to verify the location. Satisfied on this he would telephone his architect who, in turn, would telephone the planning authority to ascertain the gross building area that could be permitted, having first taken more accurate measurements of the site from the large-scale ordnance maps he kept in his office. The information the architect sought was confined to two points, the plot ratio and the permitted use of a completed building and this would be reported to his client. What followed would be an assessment of the cost of erecting the building, including fees and interest charges that would be incurred over a maximum of eighteen months. A simple calculation would provide the net area available for assessing the rental income since an allowance of 20% was the norm. A profitable deal would emerge if this income, capitalized at 10%, equalled or exceeded the addition of the development costs and site cost. Within hours of the first telephone call a deal could be struck.

The demand for offices was such that it required little or no expertise to enquire of estate agents the going rent in any particular location and builders were able and willing to quote building costs at so much per square foot. The calculations were so simple that they were literally made on a single sheet of paper and often on 'the back of an envelope' a phrase in common usage.

The financing of many of these early developments was equally simple, and commenced by the developer paying merely a 10% deposit against the purchase price of the site. The balance of the purchase price would normally be advanced by his bank who would be content with the deposit of the deeds of the land and evidence of planning permission. The loan would be limited to the period necessary to develop, with an added extension of a month or so to enable the completion of a letting.

Contractors anxious for building work would gladly enter into a contract permitting the developer to withhold any payment (including interest charges) until completion of the building works. Notwithstanding these temporary arrangements, the developer would then approach one or other of the funds and arrange for a long-term mortgage based upon the investment value of a completed and tenanted building, the amount of which would resuscitate the entire capital for which he was at risk.

The risk depended to a large extent on securing a tenant at the estimated rent on or before completion of the building. In the event such was the demand that the observed success of the pioneers in 1954 and for some years later encouraged entrepreneurs hitherto ignorant in property transactions to enter the field. The interval between buying the site and completing the building was somewhere between one year and two. What was experienced in almost every case in those early years was that prospective tenants would contract to lease before completion and furthermore rental estimates were almost always exceeded.

For the developer these were halcyon days indeed and although succeeding decades have been equally or even more profitable, in no other period could it have been so easy to transact business. Experience of taxation problems in the previous decade resulted in the retention of the investments the developer had created and as before the money power in the hands of the funds made it sensible to adopt this posture. *The Property Boom*[1] lists those men reckoned to have made a million pounds or more, albeit locked within the equity interest they held in property shares.

Apart from the ease of financing, the actual design and construction of the buildings followed a pattern universally adopted and accepted by developers and tenants alike. The architect would lead the team, employing on behalf of the developer a quantity surveyor, engineer and main contractor who in turn would employ subcontractors. The architect would prepare plans and a specification and often produce an artist's impression of what the exterior of the finished building would look like. The specification would be translated by the quantity surveyor to an approximate building cost and the contractor chosen was usually the one who had tendered the lowest.

Since the vast majority of developers were seeking the maximum profit possible the architects who were most popular were those whose designs and specifications tended to provide the maximum amount of lettable floor space, the essence of attainable maximum rent: sacrificing elaborate façades in favour of curtain wall construction and limiting floor to ceiling heights to the minimum permitted by building byelaws. Hardly a single building incorporated air-conditioning. Central heating boilers were oil-fired rather than gas fuelled and there were generally insufficient electric power outlets. In essence, the standard of building conformed to the sellers' market. Those developers and the investors who were still holding on to these buildings some 21 years later, the time coincident with the length of the tenants' leases, were to find themselves forced to improve them at prices far exceeding the original costs of construction.

The take-up of bombed sites for office and commercial development in 1954 and for some three or four years onwards gathered momentum as rents rose in response to the unsatisfied demand by would-be tenants in the City, Holborn and the West End of London. East of Aldgate Pump and north of Holborn were not popular. Although only a short distance from the City, the inner suburbs separated by the Thames saw little or no development and it was not until 1958 that developers were prepared to build in Marylebone, regarding Edgware Road as the boundary beyond which they believed tenants would not go. It would be no exaggeration to say that every building erected in EC1, EC2, EC3, EC4, WC1 WC2 W1 and NW1 secured tenants, many of whom contracted to lease before completion of building works. Rents escalated at a compounding rate on average of 8% doubling, therefore, over a period of about ten years.

The supply of centrally situated sites available for immediate development gradually receded but the idea of decentralization began to exercise the minds of the planners, who saw the problems of concentrating an increasing number of office workers in the relatively small area of inner Central London. Developers began to turn their attention to the suburbs, choosing sites adjoining railway and underground stations as obvious locations. Their plans were welcomed by the local authorities. Croydon, for example, on the Brighton line, succeeded in attracting developers by stressing the frequency of fast trains (only 12½ minutes) to Victoria. By 1963 almost every borough witnessed office development of this nature but with one or two exceptions, including Croydon, lettings were slow to achieve. The time was near when the unlet buildings were destined to prove an embarrassment or worse as the investment market observed the unloading of already tenanted buildings from the stock of those developers who needed the money to repay short-term bank loans (since without a tenant the long-term loan arrangements could not be concluded).

In central London almost all the cleared bomb-sites had been taken up and developers were turning their attention to old buildings for demolition and redevelopment. In some cases these could be the subject of repair and improvement including, for the first time since they were built, lifts and central heating. The provision of air-conditioning rarely occurred. Towards the end of the era the planners were getting more concerned with the problem of traffic congestion. Where once they insisted on generous car parking as a requisite for permission the reverse attitude was introduced: buildings capable of accommodating two hundred people were restricted to five or six car spaces; but this was not the only change in attitudes.

The encouragement offered by the planners in 1954 became by 1963 manifest discouragement. In the words of Professor Donald Denman, then occupying the chair of Land Studies at Cambridge University 'We now have only negative town planning'. Strict watch ensured that plot ratios were not exceeded. Elevation and fenestration detail, in particular, but many other constructional and architectural details were questioned or rejected. There were emerging conservationists, adverse press comments and individual objections to the proliferation of new buildings whose flat, curtain-walled elevations for all the world gave the appearance of huge cubic goldfish bowls. Press commentary at the time was predictable. On the one hand development was only a business after all. Curtain-wall buildings were the easiest and cheapest to build and if the first one to be built was considered architecturally acceptable why not others. Against this stance were the traditionalists who wanted the skyline to remain as it was, who preferred brick or perhaps stone elevations, the retention of small buildings and the tenants who occupied them. Redevelopment, in particular, threatened the enclaves of tradesmen and shopkeepers to be found in the back streets of central London which were often incorporated into the island sites that developers preferred.

Few of the available statistics relate exclusively to the period 1954—64 but a broad view of post-war office activity is plainly seen. London which lost by bombing some 9.5 million square feet out of a pre-war total of 87 million was reported by the GLC's 1964 survey to have achieved a total of 115 million by 1962 and by 1966 the total of space either completed or under construction went up to 140 million. The Stock Exchange recorded that the 1958 total of 111 quoted property companies rose to 169 by 1962. Before the war there was no special listing for property companies, such as there were came under the heading of Financial Trusts and were probably no more than twenty in number.

One reading of the statistics relative to the totals of values of ordinary shares in property companies compares March 1958 at £103 million with March 1962 at £800 million. The near £700 million difference suggests this figure is the measure of personal profit created by the developers. To quote Oliver Marriott, 'since there was no capital gains tax until 1962 this extra £700 million represented astonishing increases in personal wealth in an age which has never before witnessed the crippling effects of tax on the individual.'[1]

What had happened was a growing awareness of the immediate profits generated by development which not only increased the values of shares in existing property companies but encouraged a new generation of developers to seek Stock Exchange listing. At the same time, the institutions who previously had been content to provide finance on a simple mortgage basis now began to insist on a share in development profits by way of an equity in the development company.

Whilst the skyline of London was taking on the new limits that often reached a height of 300 feet, towns throughout the country were concentrating on new shopping centres. Many of these towns had suffered extensive bomb damage especially in the South East. Southampton and Portsmouth were typical examples where the pre-war main shopping streets were almost entirely destroyed. Developers were everywhere encouraged to carry out the schemes mostly on a leasehold basis. The local authority was satisfied with long-term fixed ground rents since few, if any, freeholders and developers alike foresaw the impact that inflation was to have.

To give some idea of the scale of the development of these new centres: Ravenseft, a subsidiary of the Land Securities Investment Trust Ltd (now Land Securities PLC), confirmed it had erected some 6000 shops and supermarkets since 1945 as outlined in the following correspondence with the author, and this was only one of the many developers in this field.

With but few exceptions the shops were immediately let, with multiple shop companies competing for prime positions. Retail trade boomed, but there were signs by 1964 that the office development sector was experiencing no such buoyancy. The October election in that year brought back a Labour administration and one of its immediate dictates was to call a virtual halt on office development (see page 188). From 4 November 1964 no building bigger than 2500 square feet was permitted for office use in London and certain other specified localities, thus abruptly ending one era and commencing a new one whose progress can scarcely have been envisaged.

RAVENSEFT PROPERTIES LIMITED

PRINCIPAL SUBSIDIARY COMPANY OF THE
LAND SECURITIES INVESTMENT TRUST LTD.

RAVENSEFT HOUSE, 60, NEW BOND STREET, LONDON, W1Y 0RP
TELEPHONE: 01-493-6070

YOUR REF.
OUR REF. CB/VB

J. Rose, Esq.,
28 Crawford Street,
London, W1H 1PL.

5th May, 1977.

Dear Jack,

I would first of all like to thank you for your hospitality last Friday and for the interesting discussion that we had.

With reference to the first paragraph of your letter of the 3rd May I regret to inform you that, having discussed the matter with my colleagues, it was agreed that we cannot divulge information which is not readily available to the shareholders.

I attach hereto a list of developments under a leasehold from, and in co-operation and partnership with the Local Authority concerned. Of these, Kilmarnock, Livingston and Warrington are still in the course of development, although Livingston is nearly complete. I also attach a list of developments without direct involvement with a Local Authority.

The total number of shops, stores and supermarkets completed is approximately 6,000.

I trust that the information I have been able to give you will be of use to you.

With kind regards,

Yours sincerely,

C. Behrens

Enc:

List of developments by Ravenseft Properties Limited and associated companies without direct involvement with a local authority

Aberdeen
Arnold
Ayr
Barnsley
Barry
Belfast
Bedford
Bridgewater
Brighton
Bradford
Brentford
Broomhill
Cambuslang
Clarkston
Croydon
Darlington
Ealing
Eastbourne
Erdington
Harrow
Kingston-Upon-Thames
Leeds
Leven
Leyland
Long Eaton
Northenden
Northwood
Northwich
Nottingham
Oldham
Perth
Pontypool
Port Glasgow
Prescot
Reading
Redhill
Swindon
Stockport
Solihull
Sheldon
Southend on Sea
Thornliebank
Wall Heath
Wallasey
Watford
Wembley
Woodford
York

List of developments under a leasehold from and in cooperation and partnership with the local authority concerned

Aldridge and Brownhills
Aycliffe
Basildon
Bath
Billingham
Birmingham
Blackpool
Bolton
Bootle
Bridgend
Bristol
Caldicot
Calverton
Cambridge
Canterbury
Cardiff
Chelmsford
Coventry
Crewe
Dundee
East Kilbride
Ellesmere Port
Eccles
Exeter
Edinburgh
Falkirk
Felling-on-Tyne
Glasgow — Gorbals
Glasgow — Castlemilk
Grays
Great Yarmouth
Huddersfield
Halesowen
Halewood
Harlow
Hull
Inverness
Irvine New Town
Keighley
Kidderminster
Kilmarnock*
Kirkby
Leek
Leith
Liverpool
Livingston New Town*
Llanrumney
London: Elephant and Castle (Southwark)
London: Notting Hill Gate (Kensington and Chelsea)
London: Stratford (Newham) (Landlords — GLC)
Maghull
Manchester Moss Side
Newbury
Newcastle-Upon-Tyne
Nuneaton
Peterlee
Plymouth
Portsmouth
Radcliffe
Rugby
Salford
Shipley
Sunderland
Scunthorpe
Sheffield
Sittingbourne
Southampton
South Shields
Swansea
Tamworth
Wakefield
Wallsend
Warrington*

*Still in the course of developmen

4.4 THE FINANCIAL BOOM 1964–73

The Brown ban which took its name from George Brown, the Minister responsible for its introduction is discussed in more detail in Chapter 5 under the heading of 'Town planning'. Its effect was to restrict the construction of office buildings in the South East but later other major cities and towns outside this area were included. Those building contracts which had been entered into before the ban came into effect were excluded. However, since these contracts were legal if both sides admitted that they were entered into verbally rather than by documents, a large number of buildings were commenced that would otherwise have waited until letting prospects got better. Fortunately for the developers would-be tenants began to take the view that the stock of empty buildings would soon run out and many of these tenants were persuaded to take larger premises than they actually needed. Furthermore, rents began to escalate due to the resuscitated demand and inflation.

From 1954 until 1964 rents had inflated at around a yearly compound rate of 8%. Over the next six years this rate had risen to some 15%, as shown in Table 4.1. Paradoxically the period from 1954 to 1964 had created fortunes for developers as the building boom was encouraged by the Government. From 1964 until 1973 the shortage caused by a Government edict to restrict building made fortunes for developers once again. For some the crash of 1973 wiped out these gains and others who survived saw a marking down of their assets (which later recovered). From 1945 onwards share prices had risen, ending in the peak in 1962, but at this point investors worried about over-development and by 1964 the property share market was off the boil. The media had begun to observe empty suburban buildings and the property market itself was busy selling good investment property fully let — the vendors being developers who were left with empty buildings which had to be financed from the proceeds of these sales. In the event almost all of these empty buildings had been let by the end of 1967.

Unless they contained a large element of office space, shopping developments were unaffected by the ban but the best of these developments had occurred by 1965 coincident with

Table 4.1 **Effects of scarcity on London office rents**

Year	*Average rent recorded PSF*	*Average rent in decimals (£)*	*Amount at 8% (£)*	*Amount at 15% (£)*
1956	20s	1.00	1.00	1.00
1957	21s	1.05	1.08	1.15
1958	22s 6d	1.125	1.17	1.32
1959	25s	1.25	1.26	1.52
1960	27s 6d	1.375	1.36	1.75
1961	30s	1.50	1.47	2.10
1962	32s 6d	1.65	1.59	2.31
1963	35s	1.75	1.71	2.66
1964	40s	2.00	1.85	3.06
1965	60s	3.00	1.99	3.52
1966	80s	4.00	2.16	4.05
1967	90s	4.50	2.33	4.65
1968	100s	5.00	2.52	5.35
1969	120s	6.00	2.72	6.15
1970	140s	7.00	2.94	7.07

the introduction in that year of the new corporation tax, one of the principal changes in the Finance Act 1965. Whereas before this time the institutions were financing by way of taking equity interests in property companies, the dividends now became virtually double taxed since corporation tax was charged on the profits of the property company and these profits going into the institutions suffered a second tax claim as the institutions' profits were distributed. Not only shopping developments but all other developments were affected and the property companies were now having to look to the lease-back system in order to obtain the necessary finance.

Between 1962 and 1966 property shares tended to remain static and below the 1962 peak, but by 1967 property shares began to perk up. As the impact of inflation drove rental values up year by year, property values began to include in their calculations a strong reversionary element and additionally property managers were insisting now on shorter rent review patterns down from the old 21 years to new seven or five years between rent reviews. It was at this juncture that the term 'reverse yield gap' entered the terminology of the property world. Until this time and with almost no exceptions property developers and investors, or dealers in existing properties, arranged their finance on the basis that the return on the investment was higher than the rate of interest paid for borrowed money. In the middle sixties it was possible to borrow money at a rate of around 8%. A common rate percentage adopted by valuers was that on completion of a building it would show a return of around 10%, thus ensuring a profit on sale. Valuers who took into account the effect of present capital value related to sharply increasing rents employed mathematical formulae that would produce a seemingly logical rate of return sometimes as low as two or three percent on the passing rent, providing that there was not too long a wait for a rent review or the end of a lease. Thus the difference between the lending rate and the purchasing rate showed a negative or reverse yield gap.

Financing the purchase under these conditions, the purchaser or developer found the answer by making an arrangement with a bank or other lending institution whereby the difference between the passing rent and the interest on the loan would be added year by year to the original loan, the total of which would be easily redeemable by the realization of the new value attributable to the property when the new escalated rent could be collected.

From 1967 onwards a selected number of property companies sought to take over other companies in the sector. The predators normally increased the nominal number of shares in their companies and used these shares in exchange for the shares of the victim company. The size of these transactions was such that although the difference between asset values and borrowings was small nevertheless the general effect was to fuel the stock market boom that had already started. It was around 1966 that the investing public saw the emergence of the property unit trusts and property bonds. These were basically managed portfolios started some by institutions others by private persons drawing huge funds into property purchases but not developments. The developers began to use these funds by way of lease-back deals. Additionally, the sheer weight of money pouring into these funds caused some of the institutions to commence purchasing properties for development and carrying out the developments themselves without using outside development companies. Some idea of the net annual investment by the financial institutions is given in Fig. 4.1 (adapted from information supplied by the Business Statistics Office). It was obvious that the competition to invest in property in every way, whether by direct investment or through property company shares, forced prices up and created a boom that lasted until 1973. The investment in property by institutions was some £89 million in 1963 and this escalated so that by 1973 they had invested a total of £633 million.

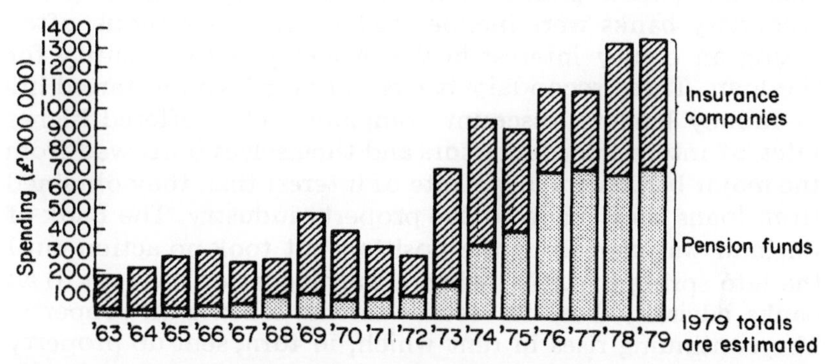

Figure 4.1 **Total institutional property investment 1963–79. The graph also shows the relative amounts invested by pension funds and insurance companies. Property spending includes both commercial and agricultural investment.**

Some three years after its institution, the Brown ban was relaxed somewhat and building (albeit on a smaller scale) was allowed through the office development permit (ODP) system. Developers who thought they had a reasonable case could apply for and obtain in many cases an ODP very much on the same lines as the building licence in the previous decade. The Conservative Government, of which Heath was Prime Minister and Barber the Chancellor of the Exchequer, believed that the recession in trade and industry could be reversed by increasing the money supply. Banks which had previously been restricted to upper limits for lending now had all restrictions virtually removed. It was erroneously believed that manufacturers would seize the opportunity to retool and otherwise expand their operations by being able to borrow almost unlimited funds. In the event, the manufacturers were hesitant to borrow heavily since it was clear to them, if not to the Government, that world trade was going into recession and that they would be wise to reduce rather than expand their capacity for output since they believed that markets were dwindling.

The business of banks is to lend money and if industry would not borrow then there were plenty of property developers and traders in property who would. The principal banks maintained their policy of lending on short terms only but secondary banks were inclined to lend on longer terms, often taking an equity interest in the property as the security for the loan. These secondary banks, many of whom started life as money lenders, discount companies, etc., offered higher rates of interest to depositors and themselves borrowed from the major banks at a lower rate of interest than they obtained from loans they made to the property industry. The Bank of England was alerted to the position but took no action until the late spring of 1972, when a warning letter was sent to all banks advising them to restrict further lending on property. The continuing rises in rent which, in turn, sent up property values were deemed to be the reason for the general economic crisis that was now apparent and it was a Conservative Government in November 1972 that decided that the remedy was to freeze rents on all business property. The repercussions of this are discussed under the heading of 'Rent control'. A small number of new buildings escaped the legislation and had the effect of creating a free market, thus new heights of rental value and, therefore, capital valuations were attained.

The inflationary spiral was further accelerated in the second half of 1973 when the consortium of oil-producing countries, mostly in the Near East and known as OPEC, increased the price of their product to four times the original price. There now emerged a crisis of confidence evidenced by the minimum lending rate rising to 13%, effectively pushing up actual borrowing rates to 18% and over. These rates made it impossible for property companies and individuals with heavy borrowings to maintain dividends or continue to value their net assets at the level of the previous year. The Heath—Barber experiment had generated inflation without real growth.

Published Government statistics show that the retail price index, one of the significant pointers to inflation, reveals a 7.9% increase in 1971, 9% in 1972, a further 7.7% in 1973 and by 1974 20%. Bank lending to industry went down by 28% in 1970 and a further 19% in 1973, whereas loans to property companies and financial institutions which had a large interest in property doubled between 1970 and 1973, the actual figures being 6½% and 13%. The money supply went from 11% in 1971 to 27% in 1973. The growth in bank lending generally went up by 19% in 1970, 29% in 1971, 31% in 1972 and 41% in 1973.

So far as homes are concerned, the Building Society Association revealed figures which showed that the price of the average house in 1970 was around £5000. By 1971 this had risen to £5650 representing a 13% uplift. By 1972 the figure was up to 30% and by 1973 35%. Thus, it could be said that the price of a house doubled in three years.

The total amount of money advanced by the building societies can be ascertained by the Government statistics published in 1981 which estimated that of the gross wealth of £755 billion of individuals, £432 billion were invested in physical assets chiefly in peoples' homes. Of the £323 billion represented by financial assets £56 billion were invested in building societies. In percentage terms this amounts to 17.6% of the total of individual wealth. The full picture is given in Table 4.2. Although predominantly concerned with lending to home buyers, their funds are not exclusively invested in mortgages as some 20% are in gilt-edged stocks and bank deposits. Those investors paying only the standard rate of tax receive net dividends after tax is deducted at lower than standard rate due to special arrangements with the Inland Revenue. The building societies thus tap the investment market of a considerable and growing number of wage earners and pensioners. The end of the financial boom occurred within weeks of 9 November 1973 when the *Financial Times*' activities property share index reported an all-time high for property share prices. By 13 November the minimum lending rate went to 13% and on the last day of that month the Stock Ex-

Table 4.2

	£ million	*% of total*
British Government securities	12350	3.8
National Savings	18593	5.8
Ordinary shares	41280	12.8
Bank deposits	41615	12.9
Building society deposits	56699	17.6
Equitable interests in life assurance and pension funds	105000	32.5
Miscellaneous	47179	14.6
	322716	

change suspended the shares of the secondary bank, London and County Securities.

Property shares tumbled down to a tenth of their 1973 high — others were suspended, never to reappear. The market for their properties became all but closed as the rent freeze threw into uncertainty the future of reversions to market rents and prices dropped on average by 30%. The financial 'bust' in property and property shares followed closely the overall equity share market. The *Financial Times'* activities all share index showed a fall of 31% during the crisis of 1973 and 1974 and finally fell by almost 60%. By December 1974 the equity market was some 27% of its peak in the summer of 1972.

Confidence in the Stock Market showed signs of renewal in 1975 when the index moved 140% upwards from its lowest peak but by then the financial establishments, including the clearing banks, the insurance companies and the pension funds, were about to make full use of the power of their money.

4.5 THE DOMINANCE OF THE INSTITUTIONS, 1973 ONWARDS

Following the property market crash the period that ensued was and indeed remains marked by the increase in direct investment in property by the institutions, who previously were content to act as lenders. The role of the developer, whether a property company or an individual, has been relegated to that of a project manager or dealer. If the latter, then finance would be by way of a short-term bank loan available either if a forward sale had been arranged or a letting that guaranteed a sale. If the former, then the institution would buy the site and provide all the development costs, giving the developer a slice of the income over and above a minimum rental income; estimated in advance this minimum would give the institution a satisfactory return. The proportions established carried with them the same ratio of future rent increases at reviews and thus the old style developers no longer enjoyed the full benefits of inflation. The institutions had at long last realized the power of the money entrusted to their care by the savings of the vast majority of the population. The path is shown in Fig. 4.2.

The crash forced many of the property companies to sell completed and let properties at bargain prices, a process that reduced their assets. According to stockbrokers Rowe and Pitman, concerning the market value of 42 quoted property shares monitored by them, the book assets dropped by stages from over £5 billion in 1974 to £4.3 billion in 1978. Yields of 5% prevailing in 1973 on prime shops and offices rose to 8% in the period between 1973 and 1976.

The property share index which on 9 November 1973 stood at 357.4 dropped to 197.96 by 20 December, some six weeks later, and slithered downwards to 79.19 by 25 November 1974. Some idea of the massive sell-out can be observed from a few examples. Marler Estates sold its prestigious Knightsbridge estate to the BP Pension Fund for £45 million; Town and City sold Berkeley Square House, the flagship of the £93 million purchase of Central and District Properties, and even the Government, through the New Town Corporations, sold off some £350 million of commercial and industrial properties. F. W. Woolworth, who previously were noted for the expansion of their chain of stores, had by 1982 sold 31 of them for a reputed £140 million.

The assets of failed property companies were first transferred to the receivers of the equally unfortunate secondary banks who, in turn, were indebted to the major clearing banks. The same receivers wisely effected an orderly disposal, taking

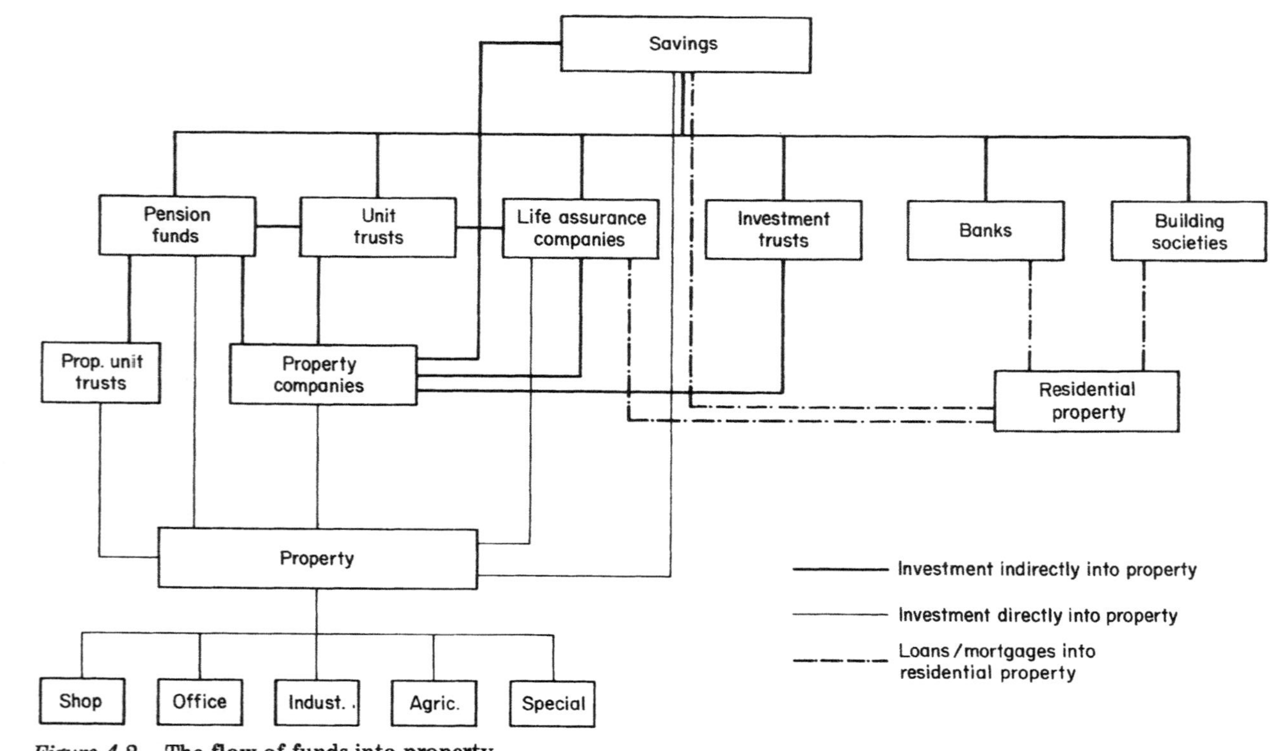

Figure 4.2 **The flow of funds into property.**

care not to flood an understandably sensitive market. The obvious buyers were the institutions, many of whom had participated in funding developments, and the Bank of England undoubtedly put pressure on the institutions and pension funds to buy since the purchases would release monies provided by the Bank to repay depositors of the secondary banks. It remains to be noted that it took some six years before the property share index reached the 1973 peak and 1979 marked a return of confidence.

By 1981 the institutions were investing directly into properties at the rate of £m1923 annually and there had been a notable decline in the committed and authorized expenditure by property companies (see Tables 4.3, 4.4).

There was one avenue of development that offered profitability with minimum tax. This was provided by the Government who introduced legislation allowing the developer to write off for tax purposes the cost of building industrial units, the purpose being to encourage manufacturing. Known as the Industrial Building Allowances, or IBAs, they encouraged a spate of new development especially adjoining motorways and airports. Much of it attracted tenants and office development of significant proportions followed, especially in the towns close to Heathrow Airport. In towns such as Reading and Slough, office rents have reached peaks beyond that attainable in certain central London districts on the fringes of the City and West End.

The building boom of the 1954—64 period made fortunes for developers for three principal reasons. The land they purchased was relatively cheap, building costs were low because they instructed their architects to provide the minimum that the tenants would accept in the sellers' market that prevailed and the tenants were made to enter into leases of 21 years in order to validate the investment valuation on the basis of the long-term finance.

When the 21 year leases came to an end in the 1980s many of the tenants vacated because of disappointment with the quality of their buildings and the availability of the alternative opportunities in what was now a buyers' market. As a result there has been a need to refurbish many of these vacated buildings at considerable cost, often exceeding the monies spent on the original construction. Additionally, there has

Table 4.3 Net annual investment by The Financial Institutions (£ millon)

		Total net acquisitions	*Net property acquisitions*	*Property as % of total acquisitions*
	Insurance	1663	307	18.5
1973	Pensions	1239	326	26.3
	Total	2902	633	21.8
	Insurance	1902	405	21.3
1974	Pensions	1441	336	23.3
	Total	3343	741	22.2
	Insurance	2509	406	16.2
1975	Pensions	3005	467	15.5
	Total	5514	873	15.8
	Insurance	3029	450	14.9
1976	Pensions	3660	631	17.2
	Total	6689	1081	16.2
	Insurance	3906	410	10.5
1977	Pensions	4559	753	16.5
	Total	8465	1163	13.7
	Insurance	4856	549	11.3
1978	Pensions	4687	702	15.0
	Total	9543	1251	13.1
	Insurance	5954	634	10.6
1979	Pensions	5563	615	11.1
	Total	11517	1249	10.8
	Insurance	6013	855	14.2
1980	Pensions	6650	967	14.5
	Total	12663	1822	14.3
	Insurance	7637	1073	14.0
1981	Pensions	6586	850	12.9
	Total	14223	1923	13.5
	Insurance	6636	1059	16.0
1982	Pensions	6753	738	10.9
	Total	13389	1797	13.4

Table 4.4 **Decline in committed and authorized expenditure by property companies 1973—79, according to information supplied by Rowe and Pitman.**

Year	*Capital commitments and authorizations (£million)*	*Interest capitalized gross (£million)*
1973	614	31
1974	700	52
1975	409	74
1976	243	73
1977	158	31
1978	169	23
1979	251	15

been considerable activity in the modernization of older buildings including those in Mayfair and similar districts where particular attention has been directed to restoring and improving facilities.

Contemplating the approach to the middle of the decade the emergence of the institutions as the dominant force in development is confirmed by statistical information. In 1971 the insurance companies and pension funds held assets of £27.6 billion:

Gilt-edge	£5.1 billion representing 18.5%
Equities	£9.27 billion representing 36.6%
Property	£3.1 billion representing 11.3%

By 1981 their total assets amounted to £138 billion:

Gilt-Edge	£32 billion representing 23%
Equities	£57 billion representing 41%
Property	£27 billion representing 19.6%

The Pension Funds whose assets account for £64 billion out of the £138 billion are forced to invest an annual inflow of £6½ billion. Fig. 4.3[7] shows the involvement in property by banks and other financial institutions, in 1982.

Not only are the individual developer and the property company in competition with the funds but they are also

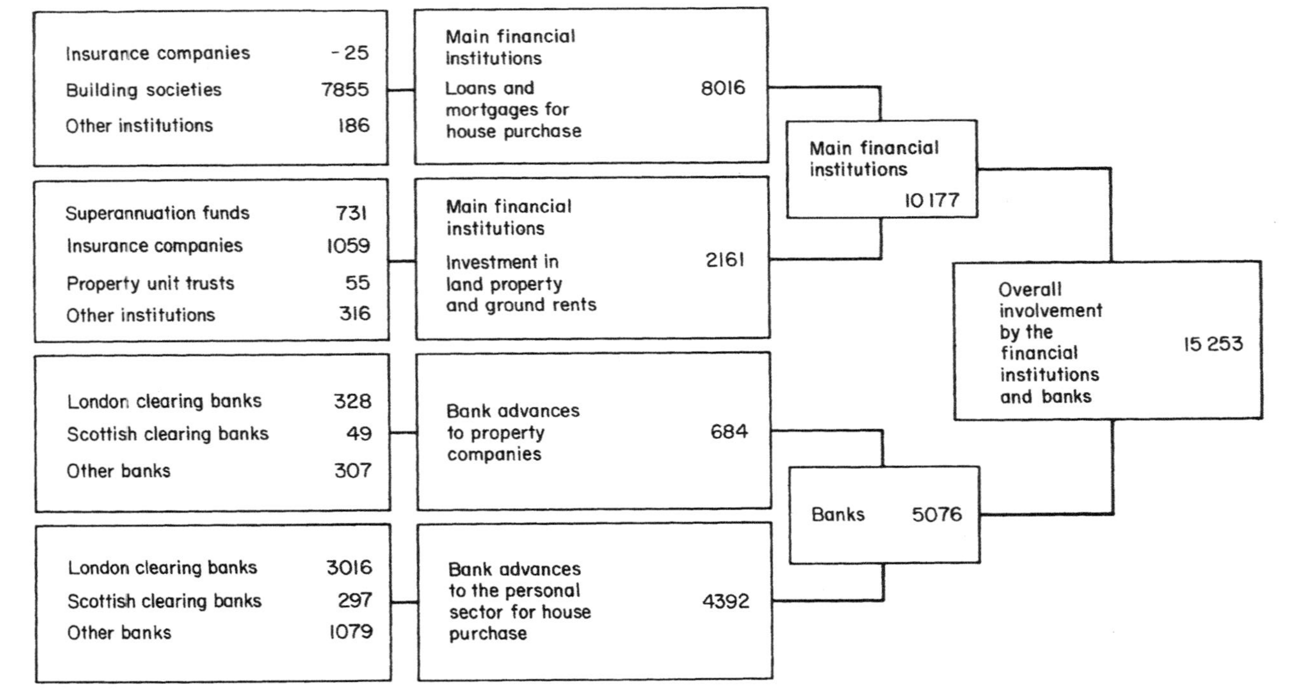

Figure 4.3 Annual involvement in property by the financial institutions and banks, 1982.

affected by local authorities anxious to attract industrial and office development in towns throughout the country who, conscious of development profit, fix financial arrangements accordingly.

In London and other major towns comprehensive schemes including leisure facilities are being funded by the institutions. They tend to favour the larger property companies as virtual project managers, though they may instead use their own staff or estate agents and surveyors employed on a fee basis.

A complete reversal of the philosophy of Labour Government policies has occurred where wholesale disposal of property controlled by central and local Government has taken the place of concepts embodied in the Town and Country Planning Act of 1947 and Community Land Act of 1975. An observer in 1974 would have been hard put to predict the vast projected development of the London Docklands to be undertaken by private enterprise. In the residential sector the curtailing by central Government of the funds necessary for local authorities to provide housing has led to a resurgence of house building by private enterprise, assisted by increasing deposits in building societies.

Unless there is a major change in legislation it is unlikely that residential property will be developed for letting. The Department of the Environment in their 1978 survey 'Attitudes to Letting' reported that 'two-thirds of private landlords were over 50 years old and well over half of all landlords were employees.' There is an active market in the sale of residential property investments, the buyers relying on obtaining possession at some indeterminate future time and then selling at an enhanced price content in the meantime with a low return. The market in existing residential property is active probably because the economic depression has caused people to re-assess their personal finances, creating a kind of musical chairs chain reaction.

Notwithstanding the seeming surplus of houses and flats for sale as is evidenced by estate agents' 'for sale' boards in practically every street, almost all could be sold immediately if the owners could buy a replacement at some lower price. There is no real surplus of housing such as occurred in the 1930s. On the other hand there is a surplus in the commercial sectors but the phenomenon that exists is the persistence of new development which tends to find tenants and at higher rents than can be obtained from existing, older properties. This trend is particularly noticeable in Central London and the so called 'golden triangles' and scientific parks near the airports and certain university campuses.

Disregarding shops which have a different basis of rental assessment, the proportion that rent represents in the total cost of a commercial concern is relatively small compared to salaries and wages. An employee earning £10 000 per annum will probably occupy as working space an area no more than 10 feet × 10 feet, which at £10 per square foot, per annum amounts to £1000 per annum. The new developments clearly offer improvements which are difficult to install in older buildings with air-conditioning, ducting and outlets for the new electronic equipment coming into use. In some cases these older buildings will be uneconomic to alter and their owners may be forced to demolish them if unable to find tenants in their existing state. Unlike the three previous periods whose commencing and finishing dates have been identified, only the commencement date of this period is clear, its end is obscured in a mist of questions, the answers to which are unknown.

Has the creation of the built environment drained too much of Britain's resources? Will the micro-chip terminal in his own home save the office worker from the need to commute and render the conventional office block redundant or enable the housewife to shop by computer link? Will the micro-chip technology applied to manufacturing processes make redundant an increasing number of manual and factory workers? How will the unemployed use their time and what development will be needed to accommodate their enforced leisure requirements? Will the private landlord in the residential sector disappear and will there be by the end of the period any market at all in residential lettings?

There are many more questions that can be asked but one final question may well interest the institutions. Will the provision of property both for residential and business purposes be left to operate a free market bereft of political interference or will its concentration in so few hands make it liable to a State take-over?

REFERENCES

1. Marriot, O. (1967) *The Property Boom*, Hamish Hamilton, London.

FURTHER READING

Aldridge, T. (1980) *Rent Control and Leasehold Enfranchisement.* 8th edn. Oyez, London.

Association of Land and Property Owners (1968) *Homes to Rent.* British Property Federation.

Balchin, P. M. and Kieve, J. L. (1977) *Urban Land Economics.* Macmillan, London.

Boddy, M. (1980) *The Building Societies.* Macmillan, London.

Charles, S. (1977) *Housing Economics.* Macmillan, London.

Davidson, A. W. and Leonard, J. E. (eds) (1976) *The Property Development Process.* College of Estate Management, Reading.

Harvey, J. (1981) *Economics of Real Property.* Macmillan, London.

5

Post-war Legislation

5.1 TOWN PLANNING

Although this section deals with post-war events, a recapitulation of the progress towards the post-war legislation helps put the subject in perspective. The origins of the relevant Acts were clearly inherent in the recommendations of the committees set up by the Coalition Governments of the two wars. There is a strong suspicion that successive governments, whatever their political colour, were caught in the trap of the legislative process. A Government in power, whether influenced by its own members or by outside pressure groups, can never be certain of the merits of any proposal to initiate or reform legislation. Presented with a problem, it has in the past, and still does, appoint a committee of experts, usually in the form of a Royal Commission, to enquire and report. The composition of the Commission cannot ever be said to be wholly unbiased in one form or another. Once the report has been completed and submitted it finds a place not only in the Cabinet Room but in the files of the Civil Service. The Ministers of the Crown, whose duty it is to act upon such reports, are unquestionably influenced by their technical advisers, usually the permanent undersecretaries in Whitehall. Richard Crossman, in his revealing *Diaries of a Cabinet Minister* writes:

> Getting briefed by Michael Stewart's officials for tomorrow's debate on Part 4, — one thing I asked is that they should give me the statements of Conservative Ministers so that we should know what they said about incomes policy. One of the civil servants replied: "I should know, because I drafted their speeches, as well as those of the present Minister." He said this in the presence of Michael Stewart, who smiled, and when I got hold of the briefs, I found that many of the words and sentences were almost identical. So it is broadly true that for five years now, the civil servants have been dictating to Ministers in successive Governments the same kind of briefs, irrespective of party.'[1]

Leaving aside the legislature, the bureaucracy has by the very nature of the beast, a propensity to expand its authority over as many facets of the activities of the population as it may. It is obvious that uncontrolled growth of the major cities without some sort of planning for the health and well-being of the inhabitants is a bad thing. Two major wars, with violent changes in the economic climate, not only of this country but world wide had undeniably produced a measure of chaos which needed some form of planning. The word itself conjured up many other aspects and under the umbrella formed by this one word sheltered not only town and country planning but the creation of a welfare State embracing Social Security and more housing, roads, schools, hospitals and other socially desirable amenities, any or all of which the legislators and pressure groups thought they could include in any planning legislation. The very first Act was entitled the 1909 Housing and Town Planning Act, but many subsequent planning Acts seemed more concerned with taxation in one form or another than with the principles of laying out towns and the countryside. Gordon Cherry's evaluation of the 1919 Act noted the growing influence on town planning by 'the growth in the professional institute, shifts in political and social attitudes in favour of more state and local intervention'.[2]

The Housing Act of 1923, the Town Planning Act of 1925 and the Local Government Act of 1929 all strengthened Government control on planning and to a very large extent followed the recommendations of the commissions and committees of the First World War. So it was with those of the Second World War. The Barlow, Uthwatt and Scott committees made recommendations that eventually appeared in the 'Control of Land Use' White Paper of 1944 and with the addition of the recommendations of the 1945 Whiskard committee the 1947 Town and Country Act was enacted. The subsequent repeal of that part of the Act which dealt with development tax left the land use control virtually untouched and led to the conclusion that the Conservative party was content with continuing State control on purely planning matters but was opposed to the Labour party's policy of appropriating the profits of development. Paradoxically the inadequacies of the taxation system persuaded the Heath Government to reintroduce special property tax provisions but there is a strong suspicion that these provisions were adopted as a sop to public clamour rather than Conservative ideology and their introduction may reasonably be considered as nothing more than political expediency. In particular, the Uthwatt committee investigated the problems of a nationalization and land allocation system. Cullingworth is quoted as saying 'under a system of well-conceived planning the resolution of competing claims are the allocation of land for the various requirements which must proceed on the basis of selecting the most suitable land for the particular purpose irrespective of the existing values.'

There is every reason to suppose that the Uthwatt committee was really searching for nothing less than a nationalization of land: An extract from the report reads:

> 'If we were to regard the problem provided by our terms of reference as an academic exercise without regard to administrative or other consequences, immediate transfer to public ownership of all land would present the logical solution; but we have no doubt that and nationalization is not practicable as an immediate measure and we reject it on that ground alone. Land nationalization is not a policy to be embarked on lightly, but it would arouse keen political controversy.'

Nevertheless, the measures which were proposed included:

1. An extension of compulsory purchase powers for local authorities.
2. The acquisition by the State of development rights in undeveloped land.
3. A betterment levy in the form of a five-yearly tax on all increases in land values.
4. The establishment of a central planning agency.

Various aspects of the Uthwatt recommendations were adopted in subsequent legislation:

1. The 1943 Town and Country Planning Act established the Ministry responsible for planning.
2. The 1943 Interim Development Act imposed a restriction on land owners' right to develop.
3. The 1943 Town and Country Planning Act gave local authorities the power to compulsorily purchase land, for reconstruction and redevelopment.
4. The 1947 Town and Country Planning Act nationalized future development rights in land and introduced a system of incremental spatial allocation dependent upon development control by local authorities. The Act also introduced a 100% betterment levy, that is a 100% tax on all increment in land value arising through no action by the owner. The State acquisition of development rights was accompanied by rights given to local authorities to buy land at existing use value and a central land board was established to administer the development charge, with powers of compulsory purchases limited, in the main, to pre-empting sales of land at above existing use value. The 1947 Act is

generally recognized as having failed to limit land sales to existing use value. Because of the control on new building, developers who did obtain building licences could afford to pay more than existing use value for land and because the levy was payable by developers rather than landowners, this caused a barrier to investment.

The Conservative party, who opposed the 1947 Act, altered it in 1951 as they were intent on raising the level of construction activity and, particularly, the rate of private house building, though within the limits of building activity set by the Labour Government. It is unlikely that the development charge procedure seriously affected the supply of land and it is probable that the Conservative Government's plans for private building would have been jeopardized by it. The Conservative party who came to power in 1951 abolished the development charge in 1953 as it constituted a permanent extra element in the cost of development. Building licences were abolished in 1954, but those parts of the 1947 Act which appertained to planning were allowed to remain 'for fear that an amendment which satisfied developers would seriously weaken or even wreck the planning machine.'[3]

The 1959 Town and Country Planning Act restored market price as the basis for local authority acquisition. Cullingworth, commenting on the Act, states

> 'The function of planning is to allocate land uses according to some desirable pattern. This is now implicitly accepted as being preferable to allocation by market forces. The retention of the illogical 1959 system had the effect of granting monopoly values to particular owners — values the very public authorities which created them had to pay for if land were needed for public purposes.'[3]

When Labour returned to power, a modified version of the development charge was reintroduced in 1967, but this time the levy was to be paid by landowners and was initially fixed at 40%. The reason for this lower rate (less than the 100% levy imposed in the 1947 Act) was to avoid stopping altogether the functioning of the market in land. Whereas the Central Land Board had disappeared, a new body, the Land Commission, was established with powers to purchase land either compulsorily or by agreement. It was given powers of compulsory purchase which would have amounted to significant curtailment of the landowner's right to transfer control of his land. Very few transactions were carried out by the Land Commission and in 1970 it was abolished by the incoming Conservative government. The land planning system reverted, therefore, to what it had been in 1959. Labour made a further attempt to tax development profits with the passing of the Community Land Act in 1975, which gave local authorities rather than central Government power to acquire land ripe for immediate or future development. The idea was for the authorities to acquire the freeholds and either develop and retain the completed building or let the land by way of building lease for a limited period with the certainty of a reversion to both the building and the land. Once again the legislation was repealed by the succeeding Conservative Government before any substantial development took place on this basis.

The problems of regionalism had never been resolved and the concept of distributing industry and commerce throughout the country, to relieve congestion in the larger cities and towns and to prevent the proliferation of distressed areas, resulted in measures such as the Town Development Act of 1952 and the green belt policy of 1960 which were economic or environmental rather than political in content. This was not the case with the Control of Office and Industrial Development Act 1965 which was an attack on office development principally in London, but later extended to other towns. The Labour party manifesto of 1964 proposed 'comprehensive regional planning directed to providing a fresh social environment'. Figures of office employment suggested growth which if continued would have added a million further office jobs in the South East and London, in particular, and it was obvious that road, rail and other services would not be able to cope. The controlling measure of 1965 could only have been politically motivated as it is difficult to imagine that the statisticians and planners advising the Government had not observed the large number of recently developed office blocks unlet and the slowing down of further development due to saturation point having been reached. Much to the chagrin of those of the Labour party who saw this allegedly planning measure as an excuse for curtailing the profits of development, two factors intervened. Firstly, although only hours separated the announcement of the ban from the moment of its implementation this period was sufficient for almost all projected developments to be contracted between developer and builder so enabling development to proceed. Secondly, faced with the curtailment of further building many would-be tenants rushed to secure what was immediately available and pushed up rental and, therefore, capital values.

Undoubtedly the most important of the post-war Acts was the 1947 Town and Country Planning Act and the comments of two of the most distinguished commentators in matters relating to planning — Gordon Cherry and Peter Hall are significant. Gordon Cherry, in his book, *The Evolution of British Town Planning*, has this to say:

> 'It is, therefore, with disappointment and humility that the planner might look back on some of his achievements. He now recognizes that any system of planned intervention in land use and social change is fraught with unknowns.'[2]

In considering the outer London Metropolitan area, Peter Hall remarked:

> 'Something like it would have happened anyway, plan or no plan. The living patterns, and to some extent the job patterns, would have occurred without a 1947 planning system at all.'[4]

5.1.1 The Whiskard committee

The Labour Government was returned to power in July of 1945 and in the following month Lewis Silkin, the Minister of Town and Country Planning, began his task of implementing the Labour election manifesto which stated 'Labour believes in land nationalization and will work towards it'. There was a White Paper on the control of land use and consideration was also given to the means of implementing the recommendations of the Uthwatt and other reports. A committee, named after its chairman, Sir Geoffrey Whiskard, was set up with terms of reference to include recommendations for the adoption of different policies, if necessary, in connection with the detail required to implement betterment and compensation. The committee made three reports in all and according to the official histories of peace-time events, *Environmental Planning*,[2] a great deal of Cabinet time was taken up before a formula was found which could form the basis of the relevant part of the 1947 Act.

5.1.2 The Town and Country Planning Act 1947

Although the Town and Country Planning Act 1947 became law in August of 1947, the operation date was fixed for 1 July 1948. This date became an important issue, because any material change in use from that which the property enjoyed on the date was liable to a betterment charge. A Central Land Board was established whose duty was to assess the appropriate amount to be levied where on the development of land an increase in value arose as a result of the grant of a planning permission.

The Act provided for a global sum of £300 million to be set aside to compensate owners for loss of any development value inherent in their properties which henceforth became vested in the State. A new development plan system brought all development under the control of counties and county boroughs, reducing the existing 1441 planning authorities to 145. Under the Act, each planning authority had to submit to the Minister a zoning scheme delineating within its area sections which it decided were to be used for housing or various kinds of other activities, including factories, shops, offices, open spaces and the like. Once the town plans had been prepared and examined at a public enquiry and finally approved by the Minister, the authority could not make any alteration without the express permission of the Minister.

The purely town planning aspects of the Act were welcomed with enthusiasm by the Government's supporters. Many of its aspects were not challenged by the Conservative party nor even by developers, probably because of the timing of the Act. 1947 was only two years away from the end of the war with Japan. A building licence was necessary before any construction whatsoever could take place and during the war no greater sum than £10 could have been spent on any property without such a licence, although this figure had been increased to a limit of £100 by 1947. The only substantial building taking place was in the repair of war-damaged properties which the district surveyor or his counterpart had pronounced dangerous structures, or in the case where a manufacturer could obtain Board of Trade support for factory alteration or construction. Property development as such was at a very low ebb and the full implication of the purely town planning aspects of the Act was, therefore, of little immediate interest to the property development world and especially so in the sector that concentrated on commercial property development.

After taking a little time to consider the full implications, property owners with little or no intention of redeveloping their holdings were soon calculating astronomical claims against the global £300 million, and the offices of the newly set up Central Land Board were inundated with such claims. Whilst there are no figures available as to the totality of the claims submitted at the time, it is probable that this bore no relation to the actual loss of development value. Fortunately for the Government, the strict control of building licences prevented any calls on the Central Land Board that were to have any significance and it was left to the Conservative government to abolish the development charge, and the Central Land Board, in the 1953 Town and Country Planning Act. The Ministry confessed

> 'many people failed to understand what the Act of 1947 had done and whether they did, or not, few landowners were willing to sell land which had development value at anywhere near its value for existing use. Whatever price they paid for land, developers saw the charge as a hindrance to development and a tax on enterprise and, of course, where they paid more than existing use value for land without any assignment of the claim and then had to pay development charge as well, the cost of development was increased.'

Almost all thinking by the Labour Government and its advisers concentrated on land for housing. In common with the rest of the planners and even the developers, they saw only dimly, if at all, the vast sums of money that in the next twenty years were to be expended on commercial development and in considering a global fund of £300 million for the loss of development rights covering the whole of the country in 1947 it comes as a sobering thought that Land Securities Investment Trust Limited was able to state in its annual report twenty years later that its assets were £1000 million. Even after allowing for inflation, the comparison of these two figures defies comment on the financial acumen of the Government of that time.

5.1.3 Legislation following the 1947 Act

The Town Development Act, passed in 1952, was intended to curtail the expansion of certain of the larger towns and cities by expanding smaller towns some distance from them. Terms such as 'decanting' and 'overspill' were added to the vocabulary of town planners. The same idea was applied to Scotland, which had its own Act, the Housing and Town Development (Scotland) Act of 1957 and a green belt policy was introduced in 1955 concerned with restricting the selection of building land.

The two Town Planning Acts of 1953 and 1954 were basically concerned with rearranging the financial aspects of the 1947 Act. The 1954 Act, however, ran into difficulties. While private sales were made at current market prices, compensation for certain planning restrictions and for compulsory purchase had to be paid on the basis of existing use plus any admitted 1947 development value. As a result, a new Town and Country Planning Act of 1959 was necessary to restore the fair market price as the basis for compensation for compulsory acquisition.

Local authorities were not only planning but housing authorities too. This meant that they had a duty to provide housing for the lower paid. Consequently, they had to compete in the market for the land they required and in broad principle, the 1947 Act would have made it possible for them to have acquired it at a much cheaper rate. Local authority housing has traditionally been more expensive to produce than private housing for a variety of reasons, the more obvious being the time taken to make decisions and, in most cases, the necessity of having to employ contractors. Most speculative house builders are financially tightly controlled and can make speedy decisions, own their own plant and employ direct labour. What would seem to be best theoretically has not, in fact, proved to be the case in practice. The consequence of the 1959 Act in restoring fair market value as the basis of compensation for compulsory acquisition, therefore, created a heavy financial burden on local authorities. The author had personal experience of these problems as a councillor and housing and town planning committee member of the Royal Borough of Kensington during the period 1963–66.

The process of planning control made little progress after the 1953 Act. The Conservative party were in power from 26 October 1951 until 16 October, 1964 and it was clearly their policy during this period to ensure as much redevelopment in all the sectors as was possible. They were, nevertheless, concerned with the recession in the economy which occurred in the middle of that period, worsening the position of the distressed areas. To alleviate the position somewhat they enacted the Local Employment Act of 1960. Amongst its provisions were powers for providing employment in localities with special dangers of unemployment and they designated as 'development districts', any locality in Great Britain in which, in the opinion of the Board of Trade, a high rate of unemployment 'exists or is to be expected'. The Act gave the Board power to provide industrialists with land or buildings in these areas, together with building grants, and special types of loans. The Board could also pay for the removal and resettlement expenses of key workers and their dependants. It set up industrial estate management corporations and, moreover, amended Part II of the Act and the regulations under Section 14 of the Town and Country Act of 1947, which dealt with development certificates.

The Town and Country Planning Act of 1962 was no more than a tidying up of the first part of the 1947 Act and dealt with development plans and the relationship between the Minister and local planning authorities and committees. There were also in this Act some amendments to planning control and a spelling out of the machinery for obtaining planning permission, together with a section of the enforcement of planning control and compensation. Special provisions applied to property owned by the National Coal Board, ecclesiastical property, and the settled land of universities and colleges.

5.1.4 Planning and market forces

The manifesto of the Labour party for the General Election of 1964 restated its policy on land control, and on 4 November 1964 stated that, having promised a system of comprehensive regional planning directed to providing a fresh social environment, it wished to check the continued growth of offices in South East of England, especially in London. It was calculated that one-third of Great Britain's population had accounted for over half of the total increase in employees over the past decade, and that about three-quarters of the South East's employment growth had occurred within the London Metropolitan Region. There was little prospect of housing adequately, inside London, more than the eight million people who lived there at that time, yet employment in London had been increasing at a rate of over 40 000 per year. Office expansion, it was stated, had been the main cause of growth and had resulted in nearly 200 000 more office jobs there since 1951. An examination of outstanding planning applications had shown that if building followed, over a million further office jobs would be created in London. It was considered that the road, railway and other services would be unable to cope with this growth and to this end the Government would shortly be introducing a Bill to control the situation. In the interim period, no building or change of use of existing buildings to office use would be allowed after midnight between 4 and 5 November, and only buildings of less than 2500 square feet, would be exempt. The Act that followed, to be known as the Control of Office and Industrial Development Act, 1965, closely followed the previous statement of intention, which made the Act less effective than it was designed to be due to the time lag.

Section 3 of Part II of the Act exempted a development where a building contract was made before 5 November 1964. The statement was made in the morning of the 4th, but was 'leaked' on the evening of the 3rd. It gave time for many developers who had not signed a building contract on land they were thinking of developing, to sign an RIBA contract with a builder or, in some cases, with a subsidiary company owned by a developer whose articles of association included the word 'builders' in it. Within a few days, a new idea was circulating through the property world. Unlike the contract for the sale of property, a builder's contract, like many others, could be considered binding if both parties were willing to state that they had indeed entered into a verbal contract at the appropriate time, and it is probable that a very large number of the planning permissions already granted were those which George Brown, the Minister responsible for the Government statement, wished to stop.

Notwithstanding the hurried rush to enter into building contracts some of which did not come to fruition or were delayed (since there was no time limit on their completion so far as the contracting parties were concerned) the Brown ban reversed a situation which had been worrying property owners for some six months or so previous to this vital date. Almost all the London boroughs were anxious to see an improvement in their rate collection and had encouraged office building with the result that around almost every tube station large office blocks had been built without finding occupiers. The rentals were at that time in the region of £1 per square foot, and £2 per square foot was the absolute maximum that could have been expected even for a top quality building. With the prospect of a ban on further office building for an indefinite period, there came within a few months a resurgence of the office letting market and these empty buildings were quickly let, at ever increasing rentals. Inflation would have had its effect on the rents that were obtained but from 1965 onwards, until the collapse of the property market in 1973, the difficulty of obtaining permission to develop office blocks in the London area had a significant effect on the rise in rentals obtained and the capital values obviously increased as well. In retrospect, the legislation could have been said to have been completely unnecessary since the market forces, left to themselves, would have resulted in a slower take-up of the vast amount of empty space existing at the time; a more orderly pattern of rent rises and corresponding capital rises would then almost certainly have occurred.

In any event, by 1964 the Ministry of Housing and Local Government found itself in difficulty, with a large backlog of planning arrears, a breakdown of the review procedure and an inability to cope with the submissions of all kinds of plans which required approval. The need for some new legislation to deal with the situation became apparent. Reports were coming in to the effect that the public was dissatisfied with the authoritative approach of the planners. Pressure groups were being mounted, and it was clear that the public wanted a say in planning. Several advisory groups and commissions were set up to report including the Planning Advisory group, the Royal Commission on Local Government and the Skeffington committee. The Government's Property Advisory Group's (PAG) report, 'The Future of Development Plans', was published in 1965 and its main recommendations were contained in the 1968 Town and Country Planning Act. The PAG report recommended a new planning procedure with two levels of plan preparation; the overall strategic policy level and the local tactical level. The Act altered the statutory framework for the preparation of structure plans and local plans and provided for a public enquiry in certain cases.

One of the more interesting clauses of the 1968 Act occurs in Section 17 of Part II, where, under the subheading of 'Established Use', it recites that 'a use of land is established if it has begun before the beginning of 1964 without planning permission in that behalf and has continued since the end of 1963.' Before the passing of this Act there had been a very lucrative trade, particularly in London, where buildings which were zoned for purposes other than offices were purchased by developers for refurbishment, and some cases where land zoned for business purposes other than offices was developed as showrooms, factories, workrooms, etc. There was no reason why, on refurbishment or even in the event of new construction, the interior finish to these buildings should not be of a very high standard indeed. This often included marble-clad entrance halls, full central heating, lino-tiled or woodblock flooring, beautifully decorated walls and ceilings, flush panel doors with modern door furniture and toilets of a very high standard. The finished effect was equivalent to, and sometimes better than many existing older office buildings. Prior to the passing of this Act, an occupier using the premises for a non-conforming use without having an enforcement order served upon him could claim an established use if he had continued that use for a period of four years. With the very high rents charged for office premises, many office-user tenants saw an opportunity to install themselves in first-class 'non-office' buildings at very much cheaper rents than in office buildings. They were conforming to planning law although in most cases paying higher than the 'correct' rents they would have been charged had they complied with the authorized use. With local authorities short of staff, neither members of the town planning departments nor the rating departments had time to make inspections and the number of enforcement notices served under the 1947 Act were consequently few.

5.1.5 Politics of the Land Commission

Although not a planning Act, the Land Commission Act of 1967 provided for the establishment of a Land Commission, giving it powers to acquire, manage and dispose of land and impose a betterment levy with provision for the payment of compensation after acquisition. In its initial stages, the Act was designed to collect the levy almost to the exclusion of all else. Later on, however, the Commission was expected to buy land which it needed, either immediately or at some indefinite time in the future. If land which had increased value because of its existing development potential was bought for public or private use, the owner would benefit by 60%, whilst the level remained at 40% of any amount attributable to betterment. If, however, land bought in advance of requirement and before its value rose on account of betterment, was later sold to a public body or private developer, the Commission would reap all the betterment. The Land Commission managed to complete the purchase of some 2800 acres of land and collect some £46 million of betterment levy before the Conservative government repealed the legislation in 1971. In short, the Act was politically acceptable to Labour and for the same reason unacceptable to the Conservatives. Its impact on planning could be said to have lengthened the process of development at its best, while adding nothing and probably subtracting something from the quality of planning and cost of development in the private sector.

One Act of Parliament that had a positive effect on planning procedure was the Local Government Act of 1972 which, in the words of Gordon E. Cherry,[2] 'may well be regarded as an important watershed in town planning affairs'. In effect it introduced a pattern of new authorities with different planning functions as between county and district councils. Some idea of the work load of the planning authorities can be given from an article in *The Chartered Surveyor* (Volume 106) in 1973. It stated that in 1971, more than 463 000 planning applications were submitted in England and Wales of which 83.3% were allowed. For the remainder, planning appeals were settled (where an inspector was appointed) in 36 weeks and, where decided by the Secretary of State, in 51 weeks.

5.1.6 The Community Land Act

The Labour party had in all its pre-election manifestos committed itself to legislate for some kind of nationalization of the land and with its return to government in 1974 set about the task, this time intending to use the local authorities as the instruments to effect it. The Community Land Act, which became law in 1975, dealt primarily with the arrangements for taking development land into public ownership. Under it local authorities were given first the power and then the duty to acquire all land needed for relevant development for the next ten years and this included all development which did not fall into the categories of exempted or excepted development.

Exempted development consisted of:

1. Development allowed under a general development order, i.e. that which is permitted by the development plan.
2. Agriculture, forestry and mining developments.

Excepted developments were of two types:

1. Defined by the nature of the development — usually so-called 'minor' developments and including industrial premises up to 1500 square metres, other developments under 1000 square metres, rebuilding for existing use if the floor space of the new structure does not exceed the old by more than 20%

2. The second type of exception is defined with reference to the state of the land, this included any development on the land which on 12 September 1974
   (i) Had planning permission;
   (ii) Was owned by an industrial undertaking;
   (iii) Was owned by a builder, or residential, or industrial developer.

5.1.7 Development land tax

The Development Land Tax Act which became law in 1976 was intended to deal with the Government's land policy in respect of restoration of increases in land value to the community. It was designed to tax profits from increases in the value of land due to development or change in existing use. Liability to the tax could have arisen when profits were realized by the disposal of an interest in land, which could be in the form of either an actual or a theoretical sale. The latter case is termed a 'deemed disposal' and a sale was deemed to have taken place immediately before a development commences. The rate of tax was set at 80%, but the first £10 000 of realized development value were not taxable. Exemption from this tax included the sale of an owner-occupied residence with less than one acre of land, or upon development of a residence for owner-occupation if the landowner was in possession of the land on 12 September 1974. There were special categories of land which were exempted from the tax, such as land held as stock in trade and with planning permission in force when first sold, leased or developed; land owned by charities on 12 September 1974 and the transfer of developments by certain housing agencies. Tax relief was also provided in certain cases as hereunder:

1. Buildings erected by industrial undertakers for their own use.
2. Buildings erected by statutory undertakers for operational purposes.
3. Buildings erected for charitable purposes on land acquired after 12 September 1974.
4. Housing built by housing associations.

On its return to power in 1977, the Conservative Government

repealed the Community Land Act but retained the development land tax.

Some thirty years had elapsed since the 1947 Town and Country Planning Act legislated by Labour for a betterment charge. It was promptly cancelled in its entirety by the succeeding Conservative administration who, when it had ample opportunity to do otherwise, allowed its retention in the modified form of the 1976 Act.

5.2 TAXATION

5.2.1 Taxes on capital

(a) Capital Gains Tax

Capital gains arising from the sale of a property were generally not subject to income tax but where the taxpayer was engaged in a sufficient number of these transactions he could find himself classified as a trader in property and his profits taxed as income. Until 1962, capital gains arising from transactions in shares were tax free. Consequently, many were carried out within the framework of a limited company, where 100% of the shares of the company owning the property were sold rather than the property itself. On this basis the property remained unsold, but the shares changed hands at a price reflecting the original cost of the property plus any increase in its value, giving a tax-free gain to the vendor of the shares. Where, however, a company was classified as a trading company, then gains were brought in as income and charged to income tax and profits tax (corporation tax). The same considerations applied to individuals where a clear distinction could be made between trading and investment.

The Finance Act of 1962 introduced for the first time a tax on short-term gains (i.e. where transactions involving both buying and reselling took place within a period of six months). In 1965, Capital Gains Tax became chargeable notwithstanding the time period between buying and reselling. As it affected property transactions, Captial Gains Tax was set at 30% usually between the sale price and cost price on a straight line basis with some relief on the proportion of the gain arising prior to 6 April 1965, or the difference between the sale price and the 1965 value with subsequent additions at cost, whichever gain is the lower. When the transaction was clearly of an investment nature and the owner not a trader in property, 30% remained the rate for tax on capital gains.

(b) **Capital Transfer Tax**

The entire concept of estate and death duties was altered by the Budget introduced by Mr Healey on the 26 March 1974, when he announced that he was proposing to introduce in the Second Finance Bill, a tax on lifetime gifts 'which will mesh with the existing estate duty so as effectively to replace it with a comprehensive tax on all transmissions of personal wealth... The new tax will take effect as from today.' Hitherto, the disposal of capital either by immediate gift or by the creation of a trust entitled the donor to reduce his estate during his lifetime and so avoid duty payable on death. Mr Healey's second Budget speech on the 12 November 1974 revealed modifications to estate duty and dealt with four stages whereby Capital Transfer Tax finally replaced estate duty. The details are basically concessionary advantages where deaths occurred between 1974 and not later than 12 March 1975 and are contained in Table 5.1[5]

In 1914 the Settlement Estate Duty Act was abolished and the 1949 Legacy and Succession Duties suffered the same fate, leaving a single death duty as the only remaining charge. The rates were, of course, altered from time to time, usually upwards, but since 1949 the increase in rates was generally accompanied by increases in the threshold, below which no duty was charged. In 1946 this figure was £100, in 1963 £5000 in 1969 £10 000 in 1971 £12 500 and in 1972 £15 000. It has been observed that although there had been increases both in rates and in yield, the significance of estate duty had declined in relative terms almost continuously. For example, in 1898 estate duty yielded £12 million and the total yield from death duties exceeded 15% of the total tax revenue. The years 1974–75 yielded estate duty of £339 million representing only 1.5% of the total tax revenue. In 1962 foreign real estate became chargeable and in 1969 discretionary trusts were also affected.

Immediately before Capital Transfer Tax replaced estate

Table 5.1 The replacement of estate duty by Capital Gains Tax

Occasion of charge	*Stage 1 before 3 March 1974*	*'The interim period'*		*Stage 4 after 12 March 1975*
		Stage 2 from 27 March 1974—12 November 1974	*Stage 3 from 13 November 1974—12 March 1975*	
Lifetime transfer	—	*CTT (Subject to ED exemptions)*	*CTT (subject to ED exemptions)*	*CTT*
Death	ED	ED Modified in one respect; the first £1000 of *inter vivos* gifts not dutiable.	ED Modified in the following further respects: (i) a lower scale of rates; (ii) exemption of property left to a surviving spouse, either absolutely or in trust; (iii) withdrawal of the 45% relief on agricultural property and business assets; (iv) withdrawal of the special relief for woodlands; (v) relief for working farmers; (vi) duty on business assets payable by instalments as before but now interest free	CTT Transitionally most ED exemptions and reliefs are preserved, including the exemption for gifts more than three years before death, the surviving spouse exemption, the reverter to disponer exemption, etc.

duty, the main categories of dutiable property were:

1. The deceased's free estate i.e. property of which the deceased was competent to dispose at the time of his death;
2. Settled property in which the deceased was entitled to a beneficial interest in possession or held on discretionary trusts under which the deceased was eligible to benefit:
3. Property given away by the deceased within the seven years before his death, or at any time if the deceased had not been entirely excluded from any benefit during those seven years;
4. The deceased's share of partnership property;
5. Property subject to an option to purchase granted otherwise than by the deceased's will to a person other than the deceased and exercisable on or by reference to his death;
6. Assets of a company controlled by five or fewer persons, where the deceased had made a transfer of property to the company, and in the seven years ending with his death, had received or been empowered to obtain benefits from the company.

The exemptions included:

1. Property in which the deceased had been interested only in a fiduciary or representative capacity by way of security (e.g. as trustee, executor or mortgagee);
2. Property passing by reason only of a bona fide purchase for full consideration in money or money's worth;
3. Property passing to:
   (i) Certain national institutions, or
   (ii) Charities up to a limit of £50 000, or
   (iii) A surviving spouse up to a limit of £15 000;
4. Settled property of which the deceased had not been competent to dispose and which had already borne estate duty on the death of the deceased's husband or wife;
5. Property which had been settled on the deceased for life and which reverted on his death to the person who had settled it;
6. Gifts *inter vivos* which:
   (i) Had formed part of the deceased's normal expenditure and had been made out of his surplus income, or

(ii) Had been made in consideration of marriage (subject to a £5000 or £1000 'ceiling' for each marriage), or
(iii) Did not exceed £500 in total to any one donee;

7. Certain Government securities if the deceased holder had been neither domiciled nor ordinarily resident in the UK;
8. Certain works of art;
9. Timber until sold.

Agricultural property and woodland held a special place in the regulations covering chargeable duty, in that only 55% of the estate rate was chargeable on agricultural property and no duty was payable on woodlands until sold. There was much business done in the purchase of these two types of assets in the closing years of a life and sometimes as near to death as the closing months or weeks of aged or seriously sick wealthy people. In the case of one typical 'off-shore' island, (The Bahamas) a realization of assets in this country could be made, transferred to that island and arrangements made there to purchase one of several estates on the understanding that the estate could be sold back to the vendors at an appropriate discount after death. In essence, the result could be likened to the client's choice of who should collect and how much -- the Inland Revenue of Great Britain or the Bahamian estate dealer.

5.2.2 Discretionary trusts

The setting up of such a Trust requires a donor plus beneficiaries and Trustees to operate the Trust. A man anxious to pass on his wealth or a large part of it after death, would use the device of a discretionary trust to escape duty by forming such a Trust which gave absolute discretion to the Trustees to deal with the capital and income within the Trust. Thus, although denied the use of the capital or income during his lifetime, the donor could keep control of the monies by appointing members of his family or close friends, as Trustees. On his death, the Trustees could at their discretion, then release to the beneficiaries (presumably the heirs) the whole of the capital. Provided that the Trust's assets had been placed there by the donor in the seven year period required to escape death duty the effect could be likened to the donor having willed his estate to his heirs without tax being applicable. The advent of Capital Transfer Tax (CTT) effectively brought this kind of operation to an end. Moreover, where there was a discretionary Trust already in being, a special provision relating to 'deemed distribution' came into force.

The effect of the legislation was to charge tax on a deemed disposal (if the Trust remained discretionary) at an ever-increasing rate at periodic intervals. There is a difference between a discretionary Trust and a settled Trust where, in the latter case, a fixed date or occasion is stipulated for the distribution of the assets to the trust. On such an occasion, the Trust is reckoned to have transferred the assets and CTT would be payable on the date that the assets are distributed. In order to make it easier for those discretionary Trusts made before CTT to be broken, special concessionary rates were introduced, and doubtless donors and Trustees anxious to avoid large in-roads of tax into the Trust, took steps before distribution to convert discretionary Trusts into settled types.

Where a Trust remains discretionary, there are five occasions of 'deemed distribution' occurring whenever an interest in possession comes into existence. Additionally, there is a 'deemed distribution' on the 10th anniversary of the date of the settlement, although it was disallowed before 1 April 1981. This periodic charge was fixed at 30% of the rates which would normally apply to capital distribution of the entire settled property. Where the Trustees of the settlement were not resident in the UK, a periodic charge was charged annually in instalments of 3% of the normal rate. Where capital distributions were made out of a discretionary Trust set up before 27 March 1974 and paid before 1 April 1980, the rates of tax were substantially reduced to make the breaking up of the discretionary Trust that much easier. The full impact of CTT is, of course, designed to break up large estates at death, by making it no longer possible to transfer large sums of money or monies worth without penal tax whenever the transfer occurs.

The basic structure of the new tax is that it is charged at progressive rates cumulatively on any chargeable transfer made by an individual during his lifetime, with death being reckoned as the occasion of the final transfer. There are two tables of rates, one applicable to transfers on or within three years of death, and the second to any other transfer. Estate or death duty in its forms up to 1894, can generally be regarded as having been of no great burden to property owners and in any event its impact had even been softened by the creation of transfers to trusts in favour of heirs or by means of gifts *inter vivos*, thus making the tax to a large extent voluntary.

There were cases, of course, where a large estate had to be sold in order to meet death duties because the owner had not taken advantage of the many loopholes in the existing law which, until 1974, had made it possible to escape the full impact of the tax. For the purposes of examining its effect upon property development, it would seem that unless future legislation reverses the present position, large fortunes will not be allowed to descend to succeeding generations. There has been a move to combat the situation however. The possession of a vast sum of money or blocks of shares produce on investment, 'investment income' which is taxed at a very high rate indeed, so much so that company directors who own large blocks of shares often forego the dividends due to them since the net amount of income left after tax is negligible. This attitude certainly existed when the top rate of income plus investment surcharge was 98%. The top equivalent rate is now 60% and we have yet to see if attitudes change.

What is normally regarded as important is the control of the company through the voting power of the shares. Where directors who own enough shares to give them voting control wish to pass this control to their heirs, it can be done in the following manner:

1. Advantage can be taken of gifts made *inter vivos* for charitable purposes, provided that such gifts are made a year before death.
2. With this provision in mind, a charitable trust can be set up, where the shares are transferred from the donor to the charity, the Trustees of the charity being the heirs of the donor.

The effect of this is such that whilst the capital and income would be the property of the charity, the donor would, at least, have the satisfaction of knowing that his heirs could control the company through the Trusteeship. There would be nothing to prevent the donor from leaving to his heirs a relatively small amount of shares upon which, of course, tax would be payable, and there would be nothing to prevent his heirs becoming directors of the company. This form of dynasty creation is likely to be more and more prevalent in the future. Chief among the charitable foundations operating on this basis are the Woolfson Foundation (Great Universal Stores) and in America, the Ford, Rockefeller and the Carnegie Foundations.

The 1974 Finance Act can be said to have had a most serious effect on the inheritance of large holdings in property itself and in property companies. Usually in times of violent economic change, the building up of large fortunes from scratch within an individual's lifetime is not uncommon, but until recent times the greater number of individuals personally holding or controlling companies with large property assets were inheritors. Clearly this latter category will no longer reign in the future. Furthermore, Capital Transfer Tax applies to any transfer at any time unless between husband and wife or within the narrow exemption limits of gifts *inter vivos.* The amounts of these gifts are in the region of a few thousand pounds per annum and can have no marked effect on the amount of tax assessable against any but a small estate.

With Capital Transfer Tax in its role as a death duty, it is inevitable that individual holdings in property will be split up. Even where the duty is chargeable against shares in property owning companies, the same result will be obtained, since if a large enough holding of shares is at stake, then the company will have to be liquidated. It is unlikely, if not impossible, that any shareholding of other than 100% in a private company can be sold. Only a controlling shareholding in a quoted public company can guarantee the company remaining intact. Most of the properties contained in such private companies, or the properties, if directly owned by the administrators of the estate of the deceased owner, should have to be sold and for the most part, these would be taken up by the insurance companies, pension funds or similar institutions.

The distinction enjoyed by a quoted public company is that it has been accepted by the council of the Stock Exchange and its shares may be quoted on the Exchange. A private company cannot enjoy this most important privilege and the sale of less than all its share capital is rendered considerably more difficult. Stockbrokers and jobbers whilst permitted to deal in private company shares, are generally unwilling to recommend clients to purchase. Brokers and jobbers have regard to the fact that the accounts of such companies are seldom publicized as is invariably the case of public companies, and the lack of daily dealings in the shares inhibits comment. The sale of such shares is often confined to those with an intimate knowledge of the company concerned, generally to directors and their families. An institution would more than likely assess the prospects of a future sale of any shares they might buy on the basis that the remaining shareholders are the best if not the only market. It is obvious that in such circumstances, an institution wishing to dispose of its shares would find itself forced to sell at a discount if at all. In Chapter 6, mention will be made of the circumstances in the early 1960s when institutions, conscious of the considerable profits being made in property development, insisted as part of any deal involving long-term mortgages that they participate in the developers' profit. Where participation was to be by issue of shares to the institutions, they insisted that the development company concerned, if not already quoted, should first obtain a quotation. In an atmosphere of continuing successful developments, many hitherto private companies took this step. Once in the hands of these public companies, properties are unlikely to be sold, being kept permanently as investments and once they are owned by corporate bodies they will presumably remain immune from Capital Transfer Tax in the forseeable future.

The Finance Act which introduced Capital Transfer Tax with taxes on gifts *inter vivos* was the one Act of Parliament which will doubtless be seen to have done more to break up large, privately owned holdings of property than any other. It follows that the future development will be more with institutions and large corporations rather than with individuals and it remains to be seen whether the same spirit in development will continue in the years to come.

5.2.3 Development Gains Tax

On 17 December 1973 the Conservative Chancellor of the Exchequer announced proposals to introduce legislation altering the basis on which tax was to be charged on capital gains arising from the disposal or letting of buildings with development value or potential. Although it was not envisaged that the legislation could be passed in the proposed 1974 Finance Act, it was the intention to tax disposals made after 17 December 1973. In the event, the Conservative Government was replaced by Labour in February 1974 and the new Chancellor included the proposals in Part 3 of the Finance Act 1974. Strictly speaking, this was not so much a new tax as a new charge, which would apply both to persons liable to income tax or companies to Corporation Tax in respect of certain gains which would otherwise have been chargeable as capital gains. Section 38 of the Finance Act 1974, provides that a development gain was to be regarded as income and as such constituting profits or gains chargeable to tax under Case VI of Schedule D. To prevent avoidance of the charge by the indirect disposal of interests in land, Section 41 of the Act closed, in advance, the loophole where the disposal took the form of the sale of shares in a land or property owning company. Section 42 dealt with the case where there was a disposal of merely an interest in land or property settlement.

One further tax liability was contained in Chapter 2 of Part 3 of the Finance Act 1974. This was the proposal, first introduced by Anthony Barber in December 1973, that the first letting of a property after development should be deemed to be a disposal of the property at its then market value for the purposes both of Capital Gains Tax and Development Gains Tax. The essence of this provision was that it gave rise to a charge to tax on the unrealized value of an interest in property. It is important to appreciate that this was the first time that a tax was levied on a profit that had not yet been made, although the Finance Act 1965 enacted that Capital Gains Tax applied to gifts.

It is interesting to note the remarks of Mr Joel Barnett explaining the intention behind the tax on first lettings. The following is an extract from his speech, as reported in *Hansard*, during the Finance Bill debate:

> 'For many years, property companies had, by a perfectly legal use of the tax laws, managed to accumulate considerable profits, to build huge empires, and had in the process been paying quite minimal amounts of tax. That

state of affairs led to a groundswell of public opinion which took the view — in so far as anyone can measure public opinion — that these profits should be taxed more heavily. . .

We are talking here about increments that have resulted from the building up of huge empires. Property companies have acquired an asset, have let it at a considerable rent, have borrowed money against the rental income and with the borrowed money have gone on and acquired more assets. As a result, not only has the rental increased but capital appreciation has taken place over a period of years.

As a result of capital appreciation on those properties, and as a result of the further increases in rental income, property companies have been able to borrow further sums of money. Those further interest charges were offset for Corporation Tax purposes and the cycle or not so much cycle as an upward spiral was seemingly never-ending. Figures have appeared in the Press, the accuracy of which I have no means of testing, but one particular investment trust was quoted as having made by way of capital appreciation and rent over seven years, no less then £500 million and paid less the £50 million tax on profits of that amount. There was nothing illegal in profits of that sort. But it is the view of the Government that the proposals that were suggested for dealing with situations of that sort last December were the very minimum that would be acceptable. That is why we have made these proposals and why I commend them to the Committee.'

5.2.4 Possible effects of taxation on the property industry

A sharp distinction must be drawn between the three main streams of property development that have occurred in the period under review and, indeed, before this. Firstly, there is development for occupation, secondly, development for resale and thirdly for investment.

It is reasonable to conclude that development for occupation escaped serious attack by way of tax, since those taxes as were imposed were often for defraying the expenses of wars, internal affairs and administration, and William Pitt's 1777 'temporary income tax' was indeed justly so described since it ceased to be levied from 1816 until 1842, when Robert Peel was forced to reintroduce it due to the economic depression. In this 'occupation' category should be included estates (predominately agricultural) which in addition to income tax on the estate's revenue were taxed further under Goschen in 1889, at a flat rate of 1% per annum on estate over £10 000 in capital value. This tax could reasonably be regarded as 'defraying expenses' rather than fulfilling political aims. It was not, of course, the first tax on capital value but followed the 1694 Stamp Act, the 1870 Legacy Duty and the 1853 and 1881 Acts, which were Stamp Acts and as such could be considered as taxes on capital value.

Property was not singled out for special taxation but the tax net, commencing with the Liberal politics of Lloyd George and the introduction of super tax in the 1909 Budget and the Finance Act of the following year, tended to embrace it; a very high proportion of the incomes affected being those derived from large property holdings. That Budget must be considered in the light not only of the economic conditions of the time but the political aspirations of the Liberal party in conflict with the House of Lords and the Conservative opposition. It is not considered necessary to relate the political history of those times but the clauses in the Finance Act relating to increment value duty, reversion duty, undeveloped land duty and mineral rights duty would appear to offer sufficient evidence of an attack on capital.

A modern Adam Smith might well now add to his four maxims a fifth, 'the tax payer should not only be taxed on his income for "defraying expenses" but also on his capital for political purposes.'

Mention has been made of the secondary purpose of wartime governments. The prime purpose is obviously to win the current war but in both the 1914–18 and the 1939–45 wars there were active committees reporting on proposals dealing with the entire spectrum of human activity. Not the least of these proposals was the redistribution of 'wealth', a word with which land owners have always been closely associated. 'Property is a sin' said Wycliffe. Marx called it theft. Whilst no great changes occurred after the 1914–18 war, the committees who reported during and immediately after the 1939—45 war were undoubtedly responsible for the 1947 Town and Country Planning Act. Its clear intention was to hold the value of land and buildings to that existing and to compensate, on a once-and-for-all-time basis, owners who would lose the right to profit by development. The profits of development were to go to the State. This Act, although later repealed (with certain exceptions related to planning), was the second attack, but this attack affected not only owner-occupiers but also developers and investors.

Finance Acts between 1945 and the present day have been used to close loopholes which allowed avoidance of tax. Included in this category have been those special provisions referred to where, for some considerable period, the sale of shares of property owning companies avoided the dealing profits being treated as income. The sophisticated way in which property finance is arranged, coupled with an unprecedented rate of inflation, made it possible in recent times to commence a development project on a rental estimate that often doubled by the end of the building operations. Doubling the rent suggests doubling the capital value and the end of permanent mortgage finance too. Developers often found themselves with enough mortgage finance to pay for both site and total development costs and even have a surplus. Tax is not payable on money borrowed. It was a Conservative Government who, in 1972, proposed legislation to tax a developer on the presumed profit he would have made if he had sold a building after first letting it. In retrospect, this move may well have been an attempt (subsequently proved abortive) to dissuade the Labour party from introducing harsher legislation when and if it came to power. This it did with the Community Land Act which once again was aimed at preventing private individuals from fully profiting from development and this fettered the hitherto free market-exchange. Clearly, both the major political parties are committed to taxing the capital inherent in land for development, with the Labour party wishing to tax wealth in any form.

What must have been a totally unexpected spin-off due to income and surtax reaching 98% at one point, was the attitude of property dealers and developers immediately following the 1939—45 War. The rate of tax charged from 1900—39 varied between 5 and 26% which was never exceeded. Until 1939, most property dealers, including contractors acting as developers, treated their transactions as trading operations. Land would be purchased and houses, blocks of shops and flats would be constructed and sold — the houses to occupiers, the shops and flats as investments after being let. The investment buyers were principally insurance companies as there were relatively few property investment companies. The risk element was always present and economic conditions did not favour the retention of property by developers when, after tax at 25% on the difference between cost and sale price, 75% of the profit came to them as tax-paid earnings which could then be used to finance further development. Assuming a total outlay of £100 000 on a development and a sale price of £150 000, the gross profit would have been £50 000, the tax payable £12500 and the net profit after tax £37 500. Inflation was minimal if not altogether absent and rents were not expected to rise above levels first obtained (often with difficulty). In many instances, sales particulars included an allowance for voids. Thus there was every incentive for a developer to sell as soon as possible and to recover the £100 000 outlay plus the profit of £37 500.

The alternative would have been to mortgage the property on completion of the development. Pre-war lending on mortgage would probably have resulted in a two-thirds advance on a valuation based on a yield of 8%. The developer would have probably based his feasibility study on a 10% return, thus throwing up an anticipated rental income on his outlay (of £100 000) of £10 000 per annum. Valued at £125 000, his maximum borrowing facility would not have exceeded £83 000. It was obvious that some five or six transactions of such a nature would have eaten up the working capital of a moderate developer although he would have created a useful investment income by the end of such a series of investment-developments.

Tax circumstances after 1945 presented an entirely new set of conditions. The same basic figures prevailed in that a satisfactory basis for development was the expectation of a 10% return on risk capital. However two important changes affected the developer: tax and inflation (although the latter was not at first fully realized). A post-war feasibility study would have resulted in the same outlay of £100 000 but with possibly a 90% tax liability on the profit of £50 000 amounting to £45 000, leaving a net of tax profit of £5000 only. To undertake a risk involving £100 000 for a net of tax profit of £5000 was clearly a business nonsense and alternative methods of arranging their financial affairs were soon found by developers.

Firstly, the development and subsequent sale of property as such ceased, except in rare instances. Individuals began to conduct their affairs by forming limited liability companies in which they held the entire shareholdings or distributed these amongst their families or associates. The subsequent sale of the shares of a company holding the property seemed the answer to the tax problem until purchasers on this basis were advised that the extraction of the property from the company would, in all probability, attract tax on the difference between the cost to the company and the increased value at disposal even to the new share owner.

The immediate post-war era was characterized by low interest rates and mortgages were readily available with rates of between 3½% and 4% common. Unlike the pre-war era, property rents (and values) began to rise. The normal period between the commencement and completion of a development project was in the order of 18 months and projects embarked upon on the basis of an estimated rent often resulted in that rent being uplifted by 50% or more at completion date. Thus an estimated rent of £10 000 per annum might well have risen to an actually obtainable £15 000 per annum and a property estimated at the commencement to achieve a value of £150 000 often became worth over £200 000. Such a value would enable the developer to obtain somewhere near £150 000 on mortgage. Even at 5% the interest payable would only amount to £7500 per annum, thus the rent would not only cover this charge but leave an income surplus of £7500 per annum and the developer, after recouping his outlay of £100 000 would be left with a capital surplus of £50 000. No tax would have been chargeable on this latter sum since it represented borrowed money, and developers were not slow in realizing the advantages of retaining rather than selling completed buildings. Chancellor Dalton, who instituted a bank rate of 2½% in 1946, made it not only cheap to borrow but, as previously explained, rising rents gave rising values and often the complete outlay was resuscitated by mortgaging and a capital surplus obtained.

Business property rents were not controlled but most house and flat rents were. It was a simple decision for many property operators to concentrate on an unfettered market. Whilst it was true that there was no control on house sale prices the profits, however large, were caught in the tax net, for 'building to sell' is dealing income. Thus the conclusion may be drawn that taxation as it affects property has in its various forms resulted in the break-up of the estates of the landed aristocracy and prevented the succession of the heirs of other large scale property owners. It has to a large extent pushed developers into commercial property development rather than residential; individual developers and property owners as such have almost disappeared, having transferred their operations to limited liability companies. In turn, these companies, in order to reduce their liability to tax, have tended to be ultimately acquired by insurance companies and similar institutions paying lower levels of tax.

Whilst this may be true of other business activities, the special tax provisions for property seem to have closed many of the loopholes in the tax structure whereby mortgaging instead of selling has allowed for considerable tax savings. The Revenue seems to have finally caught in the tax net most of the devices that enable the property manipulators to turn income into capital and escape tax on income at its highest rate. It also put an end to the practice whereby the developers paid only the lower rate chargeable on capital transactions when they finally turned their property assets into cash.

5.3 RENT CONTROL IN THE RESIDENTIAL SECTOR

By the end of the Second World War there had been a great loss of housing caused by bomb damage coupled with the fact that building licences had been introduced, restricting the amount that could be expanded on repairs and renovations. Of course, there had also been a six and a half year gap in new housing construction. Even without building licences, the construction industry could not have operated due to the severe shortage of material. In 1949, the Landlord and Tenant (Rent Control) Act imposed restrictions on premiums and the Crown Lessees (Protection of Tenants) Act of 1952 gave further protection to tenants previously excluded from the Rent Acts.

There were two further Acts in 1954. The Housing Repairs and Rent Act 1954 allowed rent increases for repairs and service charges. This Act gave exemption from control to housing associations, housing trusts and new town development corporations. It excluded from control houses newly erected or converted after 30 August 1954 whenever such work was commenced. The second Act, the Landlord and Tenant Act 1954 (Part I) extended protection to tenants under long leases at low rents. This Act applied to occupiers who were paying an annual rent less than two-thirds of the rateable value where the lease originally granted was for 21 years or more. Sub-tenants of tenants under a lease of this type also came under this legislation.

5.3.1 Creeping decontrol

The Conservative Government, elected in 1955 with a substantial majority, took a new look at the situation and on 6 July 1957 passed the Rent Act, introduced as a measure 'to remove housing from politics'. This was the first Act following the end of the war which provided for decontrol and, at the same time, set down a new basis for determining the rents of controlled dwellings.

Dwellings to be decontrolled immediately were those with rateable values exceeding £40 in London or Scotland and £30 elsewhere. The date on which the Bill was introduced was 7 November 1956 and this was the date from which the Act made decontrol effective. From 6 July 1957 all new tenancies irrespective of rateable value were decontrolled with one exception. This was when a new tenancy was granted to a sitting tenant who, before its creation, was controlled. The control or decontrol furthermore applied to the whole or part of the premises concerned and while all dwellings with rateable values above the new limits were immediately decontrolled, creeping decontrol would occur as landlords got possession. Statutory tenants and those who had leases or agreements with less than 15 months still to run were given special consideration for another 15 months and a further Act, the Landlord and Tenant (Temporary Provisions) Act 1958, extended this protection until 1 August 1960.

Previous to the 1957 Rent Act, the 1954 Landlord and Tenant Act (Part II) gave protection to mixed users, where part of a building was let for purposes of business and part for living purposes. The 1954 Act also dealt with furnished tenancies (which came under the 1946 Act) allowing their decontrol under the same limits as for unfurnished dwellings.

5.3.2 Control reimposed

The General Election of 1964 returned a Labour Government and on 17 December 1964 the Protection from Eviction Act was passed as a stop gap to prevent the recovery of possession without the order of the Court, from certain residential occupants after the expiry of a contractual tenancy below £400 in rateable value. By 18 December of the following year, the Rent Act 1965 had been passed, reimposing statutory control on houses decontrolled by the 1957 Act. The difference between this Act and the other Acts which had imposed control since 1915 was the intention of the Labour Government to make control permanent. All the previous Acts had been based on emergency conditions and had been intended to be of limited duration only. The 1965 Act replaced the word 'control' with the word 'regulation'.

Regulated rents were to be determined by a rent officer. At first this applied to dwellings which, on 23 March 1965, did not have rateable values exceeding £400 in Greater London and £200 elsewhere. Tenancies which were still controlled on 8 December 1965 were to remain subject to the rent limits imposed under the 1957 Act. It was not intended that the distinction between controlled and regulated tenancies would continue for ever, for the provisions in the 1957 Act for creeping decontrol were left to operate and it was obvious that all tenancies protected by the Rent Act would eventually become regulated tenancies. The Act contained other provisions, including additional grounds for the landlord to obtain possession but only in the case of regulated tenancies. The provisions with respect to premiums in the earlier Rent Act were amended and it was made possible for a second transmission of a statutory tenancy after death. So far as furnished lettings were concerned, the 1965 Act made the Act of 1946 permanent and for these tenancies the rateable value limits were raised to those operating under the new Rent Act.

The Leasehold Reform Act 1967 was introduced to deal with the one type of tenancy over which the 1965 Act had neglected to legislate. The Rent Act of 1957 excluded all long leases, no matter what rent was paid, from the operation of the Rent Acts. It was found that after 1965, the regulations could be evaded by landlords granting leases for more than 21 years, at rack rents. The Leasehold Reform Act 1967 provided that only leases at less than two-thirds of the rateable value would be excluded from regulation, thus repeating the position that had been established by the 1920 Act. On 8 June 1968, a Rent Act was passed to consolidate the position. This repealed and then re-enacted all the provisions of previous Acts that still applied in England and Wales, including the provisions of the Furnished Houses (Rent Control) Act 1946, with its amendments. In brief, it was an attempt to replace with no less than 118 Sections in one single Act, all Acts that had been passed between 1920 and 1957. Among its innovations was the introduction of the term 'protected tenancy' which included both controlled and regulated tenancies.

A further Act, the Housing Act 1969, commenced the process of converting controlled into regulated tenancies. It was felt that many houses still subject to control were possibly unfit for habitation or lacking in standard amenities. The 1969 Act provided for higher grants to be made by local authorities for house improvement and to give them wider powers in certain areas to bring about those improvements. Part III of the Act provided, in fact, for the conversion of controlled tenancies into regulated tenancies once they had been certified as fit for human habitation and as having all the standard amenities. The Housing Finance Act 1972 replaced Part III of the 1969 Act and provided for automatic conversion of controlled tenancies into regulated tenancies, according to rateable value, over a period of two and a half years.

The rating revaluation of 1 April 1973 was anticipated by the 1972 Act which made provision for an increase in rateable value limits. Under the Act, local authorities were ordered to operate rent rebate schemes for assisting people unable to meet their commitments. One further provision of the Act was to bring within the scope of the 'fair rent' concept housing authorities, associations and others, who had previously been exempt from the application of the Rent Acts, and who had been able to charge whatever rent they liked. Whilst these bodies were brought under the 'fair rent' umbrella, however, their tenants were excluded from security of tenure provisions.

In every local authority, a rent officer was appointed under the 1965 Act to act as a referee between a landlord and his tenant. Either could apply to this officer to determine the rent payable. The officer would normally inspect the premises concerned and fix the rent on the basis that any element of scarcity was to be deliberately ignored. Obviously, this method of assessment results in a lower rent than that which could be obtained in free market conditions. The law relating to rating allows for the valuation officer of the Inland Revenue to assume free market conditions and the anomaly is possible that in districts with high rateable values, if there is a high poundage (the rate in the pound levied by the local authority), then the rate payable could be higher than the fair rent.

Two further Acts were passed in 1974. The first, the Housing Act 1974, amended the Leasehold Reform Act of 1967 by raising the rateable value limit to £1000 for the London area and £500 elsewhere. The Act gave tenants the right to know the name and address of their landlords and provision was made so that, where landlords did not supply this, they could be brought to the Court and fined for non-disclosure. The Housing Finance Act of 1972 was also amended by the 1974 Housing Act to enable tenants to challenge accounts rendered to them by landlords for service charges. The second Act, the Rent Act 1974, brought tenants of furnished dwellings under the full protection of the Rent Act 1968, making them protected tenancies within the ambit of the rent officers. Of some importance are two types of tenancies excluded from the Act. These are tenancies granted to full-time students by educational institutions and tenancies granted for holiday purposes.

By means of the Rent Act 1974, together with those Acts from 1964 onwards, the Government had now brought under protection with security of tenure all but the most expensive property in the United Kingdom. There were, of course, some exceptions, but these were very few and for the majority of the population the free market in rented accommodation could be said to have closed down. Local authority housing was, of course, still available for those people who were high on an authority's waiting list, but most local authorities were forced to allocate the limited number of properties that were available to families who had a surfeit of children or who were living in property which was uninhabitable by reason of decay or overcrowding. For newly-married couples or those wishing to move from one district to another there was very little chance of obtaining any assistance. For many people in this position, the free market which existed in furnished tenancies provided some help until the passing of the 1974 Act. When furnished tenancies became protected, landlords were loath to create such lettings and this section of the market virtually closed down too. All that was now left in the free market was the possibility of purchasing.

Coincident with the 1974 Act came the full force of onerous economic conditions due to the oil crisis which forced interest rates to an unprecedented high level. Partly as a result of this many residential property values dropped, but at the lower and medium end of the market the fall was nothing like so severe as at the top end, nor as that which affected commercial properties. In an effort to curb inflation, the Government introduced a wages policy, and free collective bargaining by wage earners was suspended. The high interest rates combined with an insufficient drop in the price of new houses was a combination that seriously affected the ability of the average wage-earner to purchase a house. To add to the troubles of a would-be purchaser the rising interest rates had a serious effect on the finances of the building societies.

Almost all house owners purchasing for the first time borrow the major part of the purchase monies from a building society. The rate of interest the building societies charge is proportional and fairly close to the rate of interest they allow on deposits obtained from the investing public. At the time of the 1974 Act it was possible to obtain a higher rate of interest by investing in Government securities and local authority loans rather than in a building society. House building firms were already curtailing their activities at this time, and the number of houses being built by private enterprise was considerably less than in former years. The property industry, which included the builders, investors and financiers, saw little merit in adventuring into the field of building houses for sale in such a climate. To build in order to let would have been akin to financial suicide.

There has been much comment, both in this country and abroad, on the effects of rent control. The reasons or the excuses made for it and the terminology employed suggest that it is for the protection of tenants. In protecting the tenants of today, control denies the tenant of tomorrow the protection afforded by the free market. It is contended that a supply of dwellings to let equal to, or in excess of the demand would obviate the need for control. It is felt that the owner of a vacant house who does not need capital would in many cases just as soon let as sell if, by so doing the investment value he creates is equal to the vacant possession sale value. Given a climate of legislation where circumstances such as these could prevail, there would be every incentive for residential development and investment. It should not be beyond the wit of Parliament to provide for a standard form of lease for the various types of residential accommodation nor should it be impossible to convert controlled tenancies, whatever name they go under, into say three-year leases, with or without repairing covenants, at rents which are perhaps proportional in some way to rateable values.

Later, in considering the effect of rent control on business tenancies, reference will be made to Part II of the Landlord and Tenant Act 1954. This Act provides that no business tenancy shall come to an end by mere effluxion of time. A tenant and landlord, if they cannot agree amicably on terms for a new lease at the end of an existing one, must have the duration, rent and other covenants of a new lease settled for them by the Courts under the Act. So far as business tenancies are concerned, the 1954 Act has in the opinion of most landlords and tenants been a successful measure. It has allowed for the security of tenure to protect the tenant and a fair rent to protect the landlord. The measure of its success can be judged by the number of purchases of tenanted properties as investments in preference to similar vacant property.

In February 1977 the Department of the Environment published a consultation paper, the main purpose of which was to inform the public that the Government wished to concern itself with maintaining the nation's existing stock of houses, remove the complexities and obscurities of the Rent Acts and to invite evidence and comments. The scope of the review concerned itself with privately rented houses and flats and the contribution which they could and should make to meet housing needs. So far as the author is concerned, the following sentence presaged a change of direction by the Government:

> 'The Government is prepared to consider new ideas or reconsider previously rejected proposals on all aspects of the Act with exception that the general principle of security of tenure for the tenant in his home is to be maintained.'

5.3.3 Aspects of political and economic theory

So far as it goes, it seems that in the 1980s there are some signs of certain policy changes less in line with doctrinaire Socialist thinking than hitherto. Policies that merely concentrate on the present inadequate stock of houses are suspect. A policy that will result in a small surplus of accommodation could well be the aim of Governments of any persuasion and a return to a free market might seem to be an answer to the problem.

The path to this goal obviously has obstacles to overcome. Traditionally the proportion of income expended by the average tenant on accommodation under a system of control has tended to become progressively smaller as domestic rents have lagged behind inflation. Any legislation which leads to the payment of higher rents is obviously going to be resisted by controlled tenants, trade unions, politicians and voters whose sympathies are to the Left. It may be inferred from the consultation paper that the housing requirements of the nation are moving towards a two-tier system where the only choice will be between a council house or one that is owner-occupied. The private rented sector, like Cinderella, waits patiently for her Prince Charming to appear with the glass slipper of decontrol — perhaps in vain.

Long before he became a Nobel prize winner, Milton Friedman, Professor of Economics at the University of Chicago, together with his equally distinguished colleague, George J. Stigler, produced an essay on rent control entitled *Roofs and Ceilings: The Current Housing Problem.*[6] Their comments refer specifically to the USA. The essay opens with a description of the results of the San Francisco earthquake of 1906 which utterly destroyed 3400 acres of buildings in the heart of the city and where out of a population of over 400000 inhabitants, more than half lost their homes. After the earthquake, 75000 people temporarily left the city, but the remainder were accommodated in temporary camps and shelters. For many months some one-fifth of the city's former population was absorbed in the remaining half of the housing facilities left standing. In other words, each remaining house, on average, had to shelter 40% more people, 'yet when one turns to the *San Francisco Chronicle* of 24 May 1906 -- the first available issue after the earthquake, there is not a single mention of the housing shortage. The classified advertisements listed sixty-four offers, some for more than one dwelling, of flats and houses to rent and nineteen of houses for sale against five advertisements for flats or houses wanted. Then and thereafter a considerable number of all types of accommodation except hotel rooms were offered to rent.'

Friedman goes on to relate the situation 40 years afterwards when there was a nationwide housing shortage. In 1940, the population of 635000 had no shortage of housing in the sense that only 93% of dwelling units were occupied. By 1946, the population had increased by at most one-third, about 200000, while the number of dwellings had increased by about one-fifth. Therefore the city was being asked to shelter 10% more people in each dwelling unit than before the war. One might say that the shortage of 1946 was one-quarter as acute as in 1906, when each remaining dwelling unit had to shelter 40% more people than before the earthquake.

During the first five days of 1906 there were altogether four advertisements offering houses to rent as compared with 64 in one day in May of that year. In 1946 there were 30 advertisements per day by persons wanting to take a tenancy of a house, against only five in 1906. During the same period in 1946 there were some 60 advertisements per day of houses for sale as against 19 in 1906.

> 'In both 1906 and 1946, San Francisco was faced with the problem that now confronted the entire nation; how can a relatively fixed amount of housing be divided (that is rationed) among people who wished much more until new construction can fill the gap? In 1906 the rationing was done by higher rents. In 1946, the use of higher rents to ration housing had been made illegal by the imposition of rent ceilings and the rationing is by chance and favouritism.'

Friedman's view on the advantages of house rationing is that in a free market there is always some housing immediately to rent -- at all rent levels. The bidding up of rents forces some people to economize on space until new dwellings are provided, thus doubling up is the only short-term solution. The high rents act as a stimulus to new construction which eases the problem. No complex expensive and expansive machinery is necessary, the rationing is conducted quietly and impersonally through the price system.

Friedman goes on to examine objections to his arguments. If it is proposed that the rich will get all the housing and the poor none, then the facts of 1906 dispute this. At all times during the acute shortage of 1906 inexpensive flats and houses were available. The danger, if any, that the rich would get all housing was even greater in 1906 than in 1946, or after. He argues that since the income of a nation is now distributed more equally among the nation's families than before the war, then if rents were freed from legal control and left to seek their own levels, as much housing as was occupied before the war would be distributed more equally than it was then. 'It is the height of folly to commit individuals to receive unequal money — incomes — and then to take elaborate and costly measures to prevent them from using those incomes.'

> 'It is an odd way to encourage new rented construction (that is becoming a landlord) by grudging enterprising builders an attractive return. The objection to a full market in housing on the grounds that a rise in rents means an inflation, or leads to one, is met by the answer that tinkering with millions of individual prices, the rent of a house in San Francisco, the price of a steak in Chicago,

> the price of a suite in New York — means dealing clumsily and ineffectively with the symptoms and results of inflation instead of its real cause.'

One of Friedman's most telling remarks is 'As long as the shortage created by rent ceilings remains there will be a clamour for continued rent control!' At the same time the article was written, it was contended that the aggregate money income of the American public had doubled since 1940 so that the average family could afford larger and better living quarters even if rents had risen substantially.

Rents, in fact, had risen very little. They rose by less than 4% from June 1940 to September 1945, while all other items of the cost of living rose by 33%. 'The very success of OPA in regulating rents has, therefore, contributed largely to the demand for housing and hence to the shortage, for housing is cheap relatively to other things.' The essay concludes with the following sentence. 'Yet we urge the removal of rent ceilings because in our view any other solution of the housing problem involves still worse evils.'

Commenting on the situation in Paris, Bertrand D. Jouvenel tells of rent control starting with the First World War when rents were frozen. In 1922 when retail prices had trebled, rent which was normally regarded as constituting one-sixth of one's expenditure, became only one-twentieth. By 1929, retail prices had risen to six times what they had been in 1914 but rents had not even doubled. Real rents, that is in terms of buying power, were less than one-third of what they had been before the war. By 1947, laws providing for increases in rents raised them to 6.8 times the 1914 rent and this, in spite of the fact that the retail prices were by then nearly 100 times higher than in 1914. The laws were somewhat complicated, relating as they did to buildings built before certain specified dates, but owners of old buildings, (that is nine-tenths of all buildings) had been allowed to get in terms of real income, either 12% of what they got in 1939 or a little less than 7% of what they got in 1914, the law took care to specify whichever was the lesser. If a builder were now to put up flats similar to those in existence these new apartments would have to be let for prices which represent ten to thirteen times present rent ceilings in order to reward the cost of construction, or cover or show a reasonable return of capital invested. Obviously, as long as rents on existing buildings are held down artificially, it will be psychologically impossible to find customers for new dwellings at prices 10 to 12 times higher, hence construction will not be undertaken. This is the differential between the legal and economic price.

French owners are not in a financial position to keep up their buildings, let alone improve them. The very ownership of apartment buildings, far from being an asset is a distinct liability because the cost of repairs has increased from between 120 to 150 times the 1914 prices. Owners who are unable to afford repairs to the roof prefer to risk legal actions for damages in respect of the spoilation of tenants' contents. There are some 84 000 residential buildings for habitation in Paris, of which 27.2% were built before 1850 and 56.9% before 1880. Almost 90% of the total were built before the First World War. Most of the additional building was carried out immediately after that war and it slackened until 1936, when it practically stopped. It is estimated that there are about 16 000 buildings which are in such a state of disrepair that there is nothing that can be done but to pull them down. Of the remainder, 82% of Parisians have no bath or shower, more than half must go out of the lodgings to find a lavatory and a fifth do not have running water. Little more than one in six of existing buildings have been pronounced satisfactory and in good condition by the public inspector.

The French Parliament proposes to establish a new fair rent system which would not be wholly paid by the tenant, for whom there would be a special subsidy. Part of the rent would go to the owner as his return but the larger part of the fair rent would be divided into two slices. A slice to correspond with the upkeep would be paid to the owner, but not directly and into a blocked account to make sure that it was spent on repairs. A further slice for the constitution of the capital investment would not go to the owner at all, but to the national fund for building.

> 'Thus, the position of the owners would be finally sanctioned. They would be legally turned into the janitors of their own buildings, while on the basis of their disposession, a new State ownership of buildings would rear its

proud head. The judiciary verdict shows that rent control is self-perpetuating and culminates in both the physical ruin of housing and the legal dispossession of the owner. It is enough to visit the houses in Paris to reach conclusions. The havoc wrought here is not the work of the enemy but our own makers.'[7]

Closer to home, Professor S. W. Paish[8] contrasts 'the inequity of the system that discriminates between those that are lucky enough to have rent assisted houses and those that have no houses at all.' It is an economic fact that fixing maximum prices results in part of the demand going unsatisfied. Writing in 1950, he estimated that since 1919 money earnings and most prices had approximately doubled, but because rents (apart from increases in rates) had not risen at all, in real terms, the rents of some 8.5 million out of 30 million pre-war houses had been approximately halved. 'It is to be wondered that the demand for houses to let at controlled rents is enormously in excess of its supply.' He contended that the rise in the price of small houses could not be taken as linked with a rise in rents. Rising rents would follow the removal of rent restriction since the market for houses for sale is the only completely free sector of the market. This is because of the excess demand created by the artificially low rent ruling in at least two of the other sectors of the market, i.e. rent-controlled and municipal houses. His contention was that the reprieve of rent restriction would almost certainly be followed by a sharp drop in the prices of smaller houses offered for sale with vacant possession.

Professor Paish's comments on the economic effects of rent control imply much of what has been said by other writers on the subject, namely that a great strain is put on a landlord's ability and incentive to maintain the premises in good condition and further that there is a serious impediment on the mobility of labour. He considers that rent restriction is an attack on the landlord and a subsidy to the tenant, but it is a subsidy that the tenant receives only so long as he stays in his existing house. Should he leave it, he is deprived not only of the subsidy but of his right to rent another house even at the market price. It will be appreciated that tenants are consequently loath to move to new locations and that this can have a serious effect on the country's economy.

One of the by-products of rent control is the bargaining power that the sitting tenant has over his landlord should the tenant wish to purchase his home for any reason. Although Professor Paish touches on this subject briefly, in the 30 years or so since his essay there has been a growing trade in landlord — tenant sales to the advantage of the tenant. Where the tenant has either the cash available or the means to borrow it, he can usually tempt the landlord to sell to him at a price a hundred pounds more than the investment value of the house and in all probability some thousand pounds or more cheaper than the price at which the property could be sold with vacant possession. Where the tenant has the opportunity of moving to another property, whether he intends to rent or buy, he may well first purchase his present landlord's interest in order to make a profit. It can hardly be equitable for a state of affairs to exist whereby rent control can be the cause of a quasi capital tax on the landlord in addition to the actual tax he pays on his rental income.

Professor F. G. Pennance[7] criticized the 1969 Francis Committee which, reporting in 1971, offered the general view that 'the system is working well'. The system, of course, was rent control. The committee based their views on a comparison of vacancies advertised in the *London Weekly Advertiser* which showed a drop in unfurnished vacancies between 1963 and 1970 and an increase in furnished ones for the same period. Since at that time furnished homes represented virtually the only free sector of the rental market, there were obviously forces at work other than an automonous shift in consumer preferences towards owner occupation. Pennance observed that it was strange that the Francis committee forbore to draw the obvious conclusion that rent regulation had affected supply. Since the statutory definition of fair rent (at which most houses were let at the time of Professor Pennance's remarks) excludes scarcity value, it is a restricted rent (in terms of compatability) and rent officers, in determining fair rents, were inevitably determining other fair rents on the basis of those already established for comparable properties in the area. Professor Pennance termed this situation 'economic incest'. The fact that many applications to rent officers had produced increases in rent was insignificant, what mattered was by how much rent had been increased to provide investment incentive. 'Possibly there is still hope in the fact that more recent legislation has still retained the idea of housing allowances for needy renters in the private sector. Therein lies the seed of the restoration of a free market in rental housing.'

M. A. Walker, the chief economist of the Fraser Institute of Vancouver, British Columbia, looked at the opposing views of rent control.

> 'Rent control is a form of price fixing, it increases the shortage of housing and ultimately reduces the ability of tenants to choose where and under what conditions they live. Rent control is a form of tenant's protection, adopted because housing is a basic need like sunshine and fresh air, and its provision ought not to be left to the vagaries of the market place. Not surprisingly, what rent control seems to be depends on your point of view. Rent control as an aspect of social legislation cannot afford the reality that it is, in essence, a form of price control. Therefore, by definition, it creates (or exacerbates) a shortage of housing by increasing the quantity of housing being demanded and decreasing the quantity of housing supplied.'

With the possible exception of those whose politics are against private property ownership, almost all economists seem to agree that the control of rents arose from the shortage of housing and that decontrol would cause hardship to tenants. With control, of course, goes security of tenure. It is now possible that there is a case for the separation of these two concepts and that whereas the security of tenure should remain, there should now be an abandonment of rent control (a virtual tax on the landlords). Relief of any monetary hardships suffered by tenants should be undertaken by the State.

In this connection, a consideration of the provisions of the Housing Finance Act 1972 (Part 2) is of some significance. It imposed duties on housing authorities in respect of Housing Revenue Account dwellings (Section 18) and on local authorities other than the Greater London Council (which was already empowered to allow rebates) to introduce rent allowances for unfurnished accommodation in the private sector. The position in 1977 was that a payment of up to £8 per week could be allowed from public funds if need was established by the tenant. Subsequent to the 1972 Act, the Furnished Lettings (Rent Allowance) Act 1973 extended the allowances scheme to include furnished accommodation.

To summarize, the dual problems of rent control and security of tenure can be said to have resulted in legislation which, although designed to protect tenants, has proved to be like some medicines: a cure worse than the disease. Although it has protected the sitting tenant, it has denied tenancy opportunities to others. It has penalized the investor in residential property, forcing him out of the business of providing houses and deterring him from replenishing his stock. It is suggested that not until the investment value of a tenanted house equals its vacant possession sale value will the problem of housing supply begin to be solved. To remedy the situation it may help to look to the one free landlord/tenancy market that works — the commercial market — where rent control has been separated from security of tenure. Landlord and Tenant Act 1954 (Part II) type legislation with security of tenure and fair rents as between landlord and tenant would ensure the effective subsidies now being given by landlords would need to be met by the proper quarter — the State.

5.3.4 The leasehold reform act 1967

Towards the latter half of the nineteenth century the builders of housing estates, as often as not, took building leases on sites owned by large-scale landowners and then sold to individual occupiers. They apportioned to each house a ground rent of a few pounds a year. Wherever builders were able to purchase freehold land they copied the landowners by retaining a freehold interest and selling the houses on leasehold tenure, occasionally granting 999 years, but more often 99 year leases.

Rent control and tenure security legislation before 1967 applied only to those tenancies where the rent was more than two-thirds of the rateable value. The ground rent level of practically all small houses built in the last century was low enough to take them outside the existing legislation. This covered not only rent control but more importantly security of tenure.

The Leasehold Reform Act 1967 sought to redress the social and economic consequences of dispossessing thousands of leaseholders whose landlords could otherwise have obtained vacant possession on the expiry of leases imminent in 1967. The reasoning behind the measure was that the original purchaser of the long lease owned the bricks and mortar. He and his assignees should retain the building and either have the right to purchase the land on which it stood or obtain a new lease. In the latter case the freeholder could demand a ground rent based on the current value of the land.

The Act, which became effective on 1 January 1968, applied to houses but not to flats or maisonettes. To qualify for the right to enfranchise, the tenant had to prove occupation for five years (out of ten) and that the ground rent payable was less than two-thirds of the rateable value of £200, if outside, or £400 if inside London. Alternatively, the tenant could opt for an extension of his existing lease for a term not exceeding 50 years and pay a ground rent based on the present site value, this ground rent being reviewable at the twenty-fifth year. The Act was amended by Section 118 of the 1974 Housing Act which provided for new qualifying rateable value limits. Where the tenancy was created before 16 February 1966 the limit was raised from £200 to £750 outside, and from £400 to £1500 inside London. Subsequently, Section 21 of the Housing Act of 1980 reduced the qualifying occupancy time from five years to three years.

Giving security of tenure to occupiers could well have been accomplished by legislation for a continuance of occupation at market rent, controlled by the 'fair rent' system operated by rent officers and tribunals. The Act is significant in that for the first time, the process of compulsory purchase was introduced as between private persons and not between the State and the property owner.

The formula for calculating the price to be paid when enfranchising a house has been confirmed by numerous cases settled by the Lands Tribunal. In broad principle, the right to receive the original ground rent of the leasehold interest for the unexpired term of the lease constituted the first part of the valuation, to which was added the deferred value of the reversion to vacant possession of the house and land. In practice, the enfranchisement of their houses has resulted in considerable financial benefits to occupiers, while conversely reducing the freeholders' value — the latter losing the prospect of large capital gains.

The reaction to the legislation has been the creation of a market in large portfolios of ground rents at prices showing investors as much as 20% return on capital for those with distant reversions, but only a nominal return for those with only a few years to go. An application by a leaseholder to enfranchise generally occurs when he wishes to sell and finds the purchaser insisting on having the freehold either to satisfy a personal choice or because his building society is not happy with the length of the remnant of the leasehold interest. The market is active and dominated by investors who take a long-term view, carefully choosing portfolios where immediate yield from rental income is balanced by the capital appreciation expected both by the effect of inflation on vacant possession values and the rate of turnover in this class of transaction.

Apart from the abrogation of the sanctity of contract that previously operated as a cornerstone of English law, the Act somewhat illogically excluded flats and maisonettes and set arbitratory financial limits, beyond which enfranchisement could not be enforced by a tenant. If its object was to protect occupiers from eviction, this could have been achieved by introducing legislation similar to that contained in Part II of the 1954 Landlord and Tenant Act. Part II, as it applied to business tenancies, gave security of tenure but ensured that the fair market rent should be paid by a tenant irrespective of any historic rent he may have been paying prior to the end of his lease.

5.4 RENT CONTROL IN THE COMMERCIAL SECTOR

5.4.1 Historical background

The first permanent Act to apply to business premises was the Landlord and Tenant Act 1927. For some time 'mixed' users (where both a residential element and a business element existed, as in the case of a shop with a flat over it held under one lease) continued to remain controlled under the Increase of Rent and Mortgage Interest (Restriction) Act of 1920 (Section 13). Section 1 of the 1927 Act provided for compensation for a tenants's loss of goodwill if he was given notice to quit at the end of his tenancy. Section 5 of the Act gave the tenant limited rights to a new tenancy for a term of no more than 14 years, if the Courts decided the amount of the compensation for goodwill and/or tenant's improvements was inadequate. To qualify for a new lease, however, the tenant had to have carried on a business at the premises for a minimum of five years.

The legislation was not considered satisfactory since the tenant had to prove he had created a positive increase in the rental due to his goodwill or improvements. This was often a difficult task, for the tenant could seldom argue that the landlord was bound to re-let to the same trade as the outgoing tenant and that a different trader would pay a higher rent taking into account the goodwill of a trade in which he was not to follow. Professional offices, etc., were excluded from this Act.

A leasehold committee was set up in 1948 to re-examine the problem of rent control and security of tenure for business tenancies. The committee recommended an interim scheme for security of tenure but not for any form of rent control. This recommendation was not immediately acted upon but instead the Leasehold Property (Temporary Provisions) Act was passed in 1951 providing for shop tenants to apply to the Courts for a new tenancy of up to one year only.

In 1953 the Government White Paper 'Government Policy in Leasehold Property in England and Wales' (Cmmd 8713) closely followed the final report of the leasehold committee and its recommendations, and resulted in the passing of the Landlord and Tenant Act 1954, which became operative from 1 October of that year. Part II of the Act repeated many of the provisions of the 1927 and 1951 Acts; it enacted provisions for security of tenure and provided for compensation to tenants where possession could be obtained by the landlord, in certain cases specified in the Act.

These new principles, operative now for a period of some 30 years, have proved remarkably successful. The Act is probably the least contentious piece of property legislation in the history of rent control and tenure security and this is endorsed by the minimal number of Court Actions arising from its operation. The 1927 Act was principally concerned with shops but the 1954 Act included business premises of all types extending to any 'activity done for payment', this including all professional offices, clubs, doctors' surgeries, and even hospitals. It does not extend to non-resident lessees however, since physical occupancy by the tenant is necessary to bring him into the scope of the Act. Excluded from the protection of the Act are agricultural holdings, mining leases, on-licensed premises, service tenancies and premises in use for public purposes. Tenancies for a fixed term not exceeding six months are similarly excluded but periodic weekly, monthly, quarterly and yearly tenancies are not.

5.4.2 Notices and counter notices

Section 24 (i) of the Act recites that 'no tenancy to which Part II applies shall come to an end unless terminated in accordance with the Act'. Section 25 of the Act allows the landlord to serve notices to terminate the tenancy in a prescribed form (Landlord and Tenant (Notice) Regulation 1969 SI 1969 No. 177 as amended). This notice must be served on the tenant not less than six months or more than twelve months before the determination date specified in it. The tenant (unless he wishes to vacate) may serve a counter notice expressing his wish to retain possession. He may do this either under Section 25 or Section 26 of the Act, but if he chooses to do so under the latter section he, in turn, must specify the proposed terms of the new tenancy requested, and moreover must serve his notice within two months of receiving the landlord's notice.

Normally a tenant will obtain a new tenancy under the Act on terms similar to those of the old tenancy (with the exception of an alteration in the rent) for a period up to 14 years. In the event of the parties being unable to agree terms the Court may impose them, but it is limited to a maximum term of 14 years, although it has discretion on all other points in dispute, including rent.

5.4.3 The landlord's right to possession

An application for a new tenancy may be opposed if, either

in his Section 25 notice, or Section 26 (6) counter notice, the landlord can satisfy the tenant either in negotiation or in Court, that under Section 30 (1):

1. The tenant has failed to carry out repairs.
2. The tenant has persistently delayed paying rent.
3. The tenant has breached other obligations substantially.
4. The tenant has been offered other premises which in all the circumstances are a suitable alternative.
5. The landlord might expect to let more profitably the whole of which the tenant's premises forms only part, provided the multiple tenancies result from subletting and not from the landlord's actions, or may otherwise dispose of the whole.
6. The landlord wishes to demolish the whole or a substantial part of the premises and could not reasonably so do without obtaining possession.
7. The landlord (who must have owned the relevant interest on the property for five years) intends to occupy it himself for the purposes of his own business or residence.

Much of the subsequent case law has been based on the interpretation of the above provisions.

The tenant's compensation for loss of possession under 5, 6, or 7, was originally fixed at the rateable value if he had occupied up to 14 years or twice this if the period was over 14 years. A statutory instrument made on 25 March 1981 raised the compensation figures to 2¼ and 4½ times the rateable value respectively.

In discussing development requiring demolition of existing buildings, it is important to note that the provisions of Section 30 do not apply to protected residential property. Where a developer finds such a residential tenancy existing, he may well pay expensively to buy out the tenant or even be held up indefinitely if the tenant refuses to negotiate, as sometimes occurs, leaving the developer locked in. Such a situation can add enormously to the cost of a site in terms of excess interest charges over and above the lower income usually obtainable from the existing buildings on the site.

5.4.4 The operation of the Landlord and Tenant Act 1954

The market in business premises consists essentially of institutions and private sector landlords owning the majority of freeholds, or long leaseholds at ground rents, and letting to occupants at rack rentals. Such sales that take place are predominantly to investors buying existing buildings which are already let, or if vacant with the intention of letting and holding. Sites for commercial development and old buildings which can be demolished to form sites are generally purchased by developers who may either sell to investors on completion (and subsequent letting) or by people who may combine the two processes as developer/investors. The emergence of the developer-investor in large scale commercial development is a post-war phenomenon that has been examined in some detail. Suffice at this point to mention that important changes in tax legislation have made development for sale immeasurably less profitable than development for investment.

With few exceptions, and these include large undertakings with liquid funds, owner-occupiers represent a relatively small percentage of the commercial property owning sector. The principal reason for this is the lower return on invested capital that can be expected for the absolute ownership of property compared with the higher return obtainable from manufacturing, retailing, wholesaling, or operating the various 'service' industries and professions.

The development of new commercial property has occurred in such quantity and has attracted so many operators in the process that it is now regarded, especially by the participants as a new profession. The impetus, given to it by an ample supply of low interest loans in the period 1946—70, together with the expansion of industry and commerce that followed the end of the war, created a predominantly sellers' market. High demand sent rents up and competition to secure both old and new buildings continued almost without pause, although there were signs of over supply of offices in 1964. This latter situation, however, was soon reversed when in October of that year the Government introduced a rationing system for new office buildings in London, the South East and the larger provincial cities.

In the personal experience of the author who was closely involved in the development marketing of commercial property both before and after the Second World War, there existed a number of property owners who resisted the temptation to base their operations on a continuing expectation of increased demand and what that entailed, notwithstanding the euphoria that existed in the world of property and the Stock Exchange evaluation of property company shares.

In the period between the wars, commercial property had not enjoyed the post-1945 experience. Money value in purchasing terms was much more stable and far from a fear of inflation was the greater fear that as money 'got tighter' the price of goods and the rents of property would tend to fall. As a result of this, between the wars, landlords who were anxious to let vacant properties looked as much to the financial stability of the tenant as to the rent he might be induced to pay. Moreover, a tenant of undoubted covenant calibre would be encouraged to take as long a lease as possible. At that time rent revision was to all intents and purposes a concept not yet invented.

As a practical example of the importance of security of tenure for the landlord, where two identical adjoining shops existed with one let to a multiple shop tenant on a long lease at £200 per annum and the other vacant, the former would command £4000, the latter near £2000. Where a lease came to an end the landlord would in those days hardly wish to evict his tenant if he could negotiate fresh terms at a reasonable rent and only special circumstances would make the landlord wish to obtain possession, such as is now provided for in the 1954 Act.

Since the 'oil crisis' and the subsequent recession in the commercial property market of the 1973–75 era, there has been a tendency to return to pre-war attitudes on rents and security of tenure. In addition to the loss of rent occasioned by the tenant's acceptance of a notice to quit served under the 1954 Act, there is also the question of compensation which may be due to him. Rateable values, on which compensation is based, have risen considerable in the period 1954–75 and are now so close to rental values as to be indistinguishable in city centres.

If a post-war office building of some 20 000 square feet in area is situated in one of the more favoured thoroughfares of London, it may well have a rental value of some £140 000 per annum (£7 per square foot). The rateable value will be close to this and the compensation for a tenant of 14 years standing is likely to be 4½ times this figure, in the order of £630000. To go further: if the landlord obtains possession and is then unable to let, he becomes liable to pay rates, probably in the region of some 100p in the £1. Furthermore, Section 16 of the Local Government Act 1974 could make him liable to double or more of this amount if the premises remain unlet.

In addition, there could be a situation arising that would add further costs to the landlord. If the tenant had, during his occupancy, carried out improvements, he might well be entitled to compensation for these. Having obtained possession, the landlord may not be able to re-let immediately. He would become liable for void rates or even 'penal rates' and his annual rate bill in the example stated would probably exceed £100 000 per annum. Such considerations would undoubtedly weigh heavily on the mind of any landlord seeking to obtain vacant possession as the alternative to re-negotiating terms with the sitting tenant.

Part II of the 1954 Act with some justification, can be regarded as a wholesome piece of legislation that offers equitable remedies against most of the injustices landlords and tenants could inflict on each other. It provides an object lesson to legislators so far unable to solve the difficulties of the residential section. Its most important ingredient is that it separates rent control from security of tenure, giving the latter to the tenant, thus allowing him to identify himself virtually permanently with the property; the ultimate protection of the Courts with regard to rent, however, prevents the tenant from, in effect, obtaining a subsidy from his landlord or alternatively from being held to ransom. The landlord similarly protected has the means to maintain the property in a proper state of repair even if this liability is passed on to the tenant, since the rent will be adjusted to give effect to this increased liability of the latter. Where really necessary the commercial landlord is able to obtain possession, unlike his residential counterpart; thus redevelopment, a desirable ingredient for the property industry (and indeed for the environment) is possible. Where a landlord obtains possession, the tenant is compensated in accordance with provisions laid down by the legislation.

5.5 THE RENT FREEZE, 5 NOVEMBER 1972–19 MARCH 1975

Although forming only a part of the package of measures taken out to hold incomes, prices and rents at a standstill, the Counter-Inflation (Business Rents) Order 1973 came into operation on 29 April 1973, replacing an earlier Order of 5 November 1972. The reasons for its imposition and its perhaps unexpected effects, not only on the world of property, but indeed the entire banking and financial system of the country, need both an explanation and a commentary.

A Conservative Government took office in June 1970 and the Chancellor of the Exchequer, Anthony Barber, set up a working party chaired by Lord Crowther to examine and report on the advisability of increasing the money supply so that credit should be readily available for capital investment by industry. The report was issued in 1971, it recommended competition credit control (CCP) and the Chancellor announced the relaxation of hitherto closely restricted lending facilities by the banks to reflate the economy. As a direct result of these measures the money supply as recorded by official figures published by the Treasury increased from the middle of 1971 to the middle of 1972 by 25%. As is normal in such circumstances, some of this increase found its way into equity shares quoted on the Stock Exchange and an examination of equity prices to the early part of 1972 shows the upward movement to be expected.

The Stock Exchange, according to stockbrokers of long experience, will generally react favourably to slight or moderate inflation, but is sensitive to hyper-inflation. It was soon realized that wage increases and other cost increases that resulted as inflation progressed would seriously affect profits and the Stock Market took a sharp downward turn. Investors who had lost money lost faith too. Many turned to property investment. The industrialists for whom the easing of credit was intended took very little of the readily available money. In August 1972 the Bank of England noted from other banking reports that the expected rise in borrowing from industrialists was only 3% whilst lending to property interests rose by 50%.

During the period up to 1971, there had been a general recession in world trade and UK manufacturers were of the opinion that 1971–72 was not a propitious time for expansion, retooling or rebuilding factories. In August 1972, the Bank of England circulated a letter to the banks requesting them to cut back on loans to property companies and to concentrate on industrialists. If a conclusion could be drawn from the events to this point, it is that the Government and the Bank of England had made the classic mistake of bringing the horse to the water only to find it could not be made to drink. The supply of credit did not govern the demand. The Bank of England's letter was largely ignored by the banks. For one thing they were almost certainly fully committed in advance to a large number of loans on uncompleted property purchases, and for another they saw no reason why hitherto safe and profitable business should not be continued. No further action by the Bank of England or the Government took place during the period from August 1972 until November of that year, since the Government was still convinced that the only way of encouraging industry to expand was by continuing to provide credit for investment. This policy, in spite of advice to abandon it, continued through into 1973 until, in November of that year, various pressures forced a reversal. By November 1973, the balance of payments situation had considerably worsened and inflation increased.

Property has traditionally been regarded as the one sure hedge against inflation and the effect of monetary inflation had the inevitable result of accelerating rent increases in the free market of commercial property and this resulted in proportionally increased capital values. Few, if any, property owners contemplated a future where voids might have an effect both on income and capital value, although this is what eventually happened. This then was the atmosphere in which the property world found itself at the date of the freeze, but to complete the economic picture, mention should be made of continuing inflation, the worsening of the balance of payments situation and the Yom Kippur War of September 1973. A few weeks after the cease-fire of this war, the Arab World, unable to defeat Israel, sustained as she was by the western democracies, retaliated by quadrupling the price of oil sold through the Organization of Petroleum Exporting

Countries (OPEC). By November 1973, the British Government was in no position to continue its reflation policies. The minimum lending rate was fixed at 13% and bank lending was severely restricted. It may be recorded that a week or so before this occurred the first of the 'secondary banks', London and County Securities, collapsed.

As there was a Conservative Government in power, the business community at first accepted the freeze on rents as one of the necessary measures to contain inflation. The party in opposition and the trade unions were of the opinion that the freeze was the least that could be done in the circumstances, although later the unions were among the most vociferous in demanding its end when they saw the effect on their pension funds' investments in property. What had been generally overlooked was the vast investment in property by the pension funds of money provided by millions of union members whose pensions were threatened if the freeze were to continue for any length of time. Yet another factor opposing the rent freeze was the relationship of the clearing banks, the 'secondary banks' and some of the more speculative property companies and individuals. Since 1945, there had been growing interest in property investment as the gradual but obvious decline in the purchasing power of currency seemed to be counter-balanced by the increase in rents obtainable from commercial property. Not only were insurance companies and institutions diverting larger and larger percentages of their funds into property, but banks (especially the secondary banks) were content to lend generously to less secure organizations often without completely independent valuations.

The number of public property companies quoted on the Stock Exchange had more than quadrupled since 1945, the public had grown used to publicity on matters concerning property and the personalities connected with the many multi-million pound schemes involved. Property in 1973, was enjoying an ever-increasing reputation. In the period commencing with the removal of building licences in November 1954 until November 1964 the *Investors Chronicle* estimated that office rents in central London had risen on average at a compound rate of 8% annually. The imposition of building controls affecting development in 1964 altered this pattern and from November 1954 to the date of the article, January 1971, it was found that the rate had increased to 15% per annum. Whether this was entirely due to a shortage of accommodation or in part to the general inflationary tendency is a matter of conjecture, but the article warned that . . .

> 'If a conclusion can be drawn, it is that an artificial shortage has produced an artificial price for an article which is capable of being produced and which, in time, undoubtedly will be produced. Investors may remember the situation in the car market immediately after the last War when a new car was virtually unobtainable and a vehicle listed by the manufacturer at say £300 would change hands at up to £150 more in the black market that sprang up (while a similar but second-hand pre-war car might fetch more than the manufacturer's price for the new article). When car production began to catch up with demand the fall in prices was both sudden and drastic. The present Government is basically anti-doctrinaire in its attitude towards those who need space and those who produce space. It cannot be too long before the supply/demand ratio existing before 1964 is re-introduced and at that point office rents will stabilize. Rents will, of course, continue to rise in step with the rate of general inflation, but not necessarily taking the present artificially inflated base as a starting point. It could well be at a point towards the end of the life of the present Government that we start thinking of minus signs rather than plus signs when talking of reversions.'

5.5.1 Reverse yield gap

For many property speculators the warning was unheeded since, in their opinion, there was the clearest evidence that whatever rent would be obtained in January would be exceeded in no small measure by the following December. Where properties could be purchased with reversions or rent reviews in the short term, possibly up to seven years, present yields were virtually ignored and prices were paid reflecting 3.75–4.5% on present receivable rents, even if those same rents reflected the full rental value at the time of purchase.

Not unnaturally, the financing of such purchases was costly. In most cases clearing banks, unless supported by additional collateral securities, did not entertain business of this kind. The prospective customers had to approach or were possibly directed to one of the more adventurous finance houses prepared to lend. These lenders were often organizations who had originally commenced in the business of financing hire purchase or who had discounting arrangements with retailer customers. In this type of business the companies collected the debt over a period and charged the customer interest based on the original advance. A further change from this type of operation to property mortgage transactions was seen as a normal extension of a business which, although usually bearing a title akin to 'finance company' or 'credit company', was considered by its customers to be carrying out the functions of a bank. These establishments did, in fact, become known as 'secondary banks' although their terms of business were, of course, more onerous than those of the clearing banks. Interest rates were at least 2–3% higher and they charged a fixed sum for advancing the loan which was added to the indebtedness of the borrower.

Because they were aware of the borrower's intentions, these bodies often included a provision for a share in any profit arising on the resale of the mortgaged property and early successes in this direction encouraged them to advertize for customers for this type of business. In time the 'secondary banks' became heavy borrowers from the clearing banks but at normal interest rates. The clearing banks had no reason in the early stages to doubt the financial stability or acumen of the secondary sector since in most cases interest charges were met by receipts from the allied activities of hire purchase and similar operations.

There remained the problem of the reverse yield gap. Clearly if a property let at £10 000 per annum was bought to show a yield of 4%, it would cost £250 000. If this sum was then borrowed at 10% the annual interest charge would be £25 000 per annum and a deficit of £15 000 (6%) would arise. This was met by a method which came to be known as 'rolling up the interest'. In other words, the interest was added to the outstanding capital year by year until redemption date, and each additional tranche of interest would itself then rank as a loan to command interest. The example quoted is not outside the experience of the author, but even if the original purchase price had been on a 6% basis with reversion deferred seven years, such financing arrangements would have necessitated an anticipation of a rent increase of proportions that only hyper-inflation could achieve.

The effect of the rent freeze of 1972 was, therefore, catastrophic for all those caught in this gap. There were many leases whose rent reviews or renewals were due within a few weeks or months after November 1972, and large sums advanced by the secondary banks were due for repayment following anticipated rent increases. Banks of all types work on a revolving credit system whereby firm forward lending commitments are covered by arranging short-term borrowings from other banks. In the event secondary banks, pressed to pay short-term borrowing, called on their customers to repay capital debts which the latter had expected to redeem following the sale of their properties with agreed rent reviews. The expected rent increases were now not able to be implemented and the property borrowers defaulted on their loans. The secondary banks followed.

As in some cases the secondary banks had obtained permission to accept deposits of money from small investors under Section 123 of the Money Lenders Act, it was essential to protect the depositors. For these reasons, the clearing banks, together with the Bank of England, launched a 'lifeboat'. Until completion of the complicated task of assessing assets and liabilities of all concerned with these 'deposit' secondary banks, monies were made available to them so that no depositor suffered. To have done otherwise may have precipated a general run on all banks with a possible civil upheaval ensuing; fortunately because of the action taken, this at no time seemed likely to occur. Many of the secondary banks went into receivership, as did a number of property companies and individuals. As a result two important effects were felt. First the sheer quantity of property of all descriptions for sale by the recievers forced yields upwards. Prime office and shop properties that had been difficult to obtain at 4.75% touched 6% and even 6.5%. Secondary properties varied from 8 to 14% according to their age and the standing of their tenants. Third-rate properties such as multi-tenanted workrooms and warehouses and back street shops of doubtful age which previously had some sort of a market at 10—11% now became either unsellable or reached 20% or more.

Hardest hit of all in real terms were properties of a reversionary nature. In spite of requests, the Government had given no indication as to the duration of the squeeze, or if it was to be amended or lifted gradually. In the absence of this information, investors simply took the view that a reversion date could not be fixed and a property could not, therefore, be valued on the basis of a calculation taking into account time and rental value. There also came the realization that amongst the largest investors in property were the pension funds of the unions and staff of countless institutions and commercial concerns throughout the country. Whilst there is no evidence to support this contention, it would be naive not to suppose this fact was one of the reasons why a Labour Government repealed a Conservative Government experiment on rent control.

In view of the many difficulties the freeze caused, the Bank of England, together with the clearing banks, brought pressure on the Treasury and the Department of the Environment to bring it to an end. They hoped, thereby, that the vast amounts of their money locked in the property market could then be recouped. The 'lifeboat' at the most crucial period ran into £1200 million. An example of the need for this assistance was that one secondary bank, First National Finance Corporation Limited, borrowed £280 million. The lifeboat was originally designed to prevent a general run on all banks and it was thought that the higher rates offered to depositors by secondary banks would bring their deposits back from the clearing banks to whom many had transferred following the failure of several of the former.

It took some time to become apparent that the property assets which the secondary banks had accepted as security for loans had depreciated as property yields followed rising equity yields in the general economic slide downwards. The lifeboat exercise originally intended as a recycling operation has yet to be assessed in terms of profit and loss. If, during the period of commercial rent control, those tenants who escaped paying the fair market rent for their premises maintained the prices of the goods or services they sold, there may have been some excuse for the imposition of the rent freeze. Undoubtedly, if price maintenance had been held, any tenant so doing would have sought and obtained the publicity such a sacrifice would have deserved. The lack of evidence of this having happened must lead to the obvious conclusion. The far-reaching effects of rent control in this particular instance were not entirely due to the control *per se.* The supply and demand situation prior to 1973 was already in a state of imbalance and it would be unwise to ascribe to the freeze alone the collapse of the secondary banks and the more speculative element of the property industry.

The financial crises and the depressive effect of world trade (and, therefore, trade in the United Kingdom) would inevitably have caused a lessening of demand for commercial space and a consequential halt in rental escalation. The period 1972–75 could be likened to an experiment in a chemical laboratory. The ingredients were already simmering in the retort. The solution of market forces was already at work, transforming those substances into the new form they would ultimately take. The addition of the unwanted catalyst of rent freeze came very close to exploding the entire apparatus.

The freeze came to an end with the Counter-Inflation (Business Rents) Decontrol Order No. 21, which came into operation on 1 February 1975. Control was actually removed on 19 March of that year. The freeze imposed by a Conservative government was thawed out by Labour, who came to power following a General Election called by the Conservative Government before the end of its five-year term. It may be said that the failure of its economic policies and its handling of industrial unrest, in more-or-less equal proportions, swung public opinion to the left.

Whatever comment is valid to the Government's handling of the economy, it is patent that interference in the free operation of supply and demand of the commercial property market produced a crisis with effects reaching into the entire world of finance. If nothing else is considered, the freeze has with some certainty confirmed the role that commercial property plays in the stability of the economy provided that market forces are allowed to control it. Whilst further analysis of the subject matter of this chapter is contained in Chapter 6, a political thread is clearly visible in the fabric of rent control and security of tenure. Although the sectors of residential and business premises have divided, it is clear that there is no easy solution to the reality of an unsatisfactory landlord and tenant relationship. Political influence had been avoided in the commercial sector until what can only be described as instant politics imposed a rent freeze, the result of which, unless quickly forgotten, should deter further political action in this sphere.

5.5.2 The effect of the rent freeze on a major property company — a case study

In January 1977, the President of Town and City Properties Limited, who a few months earlier had retired from his position as chairman and chief executive of that company, gave the author the opportunity of a personal interview to discuss the fortunes of that company from its inception to its amazing expansion; the policies he adopted and the misfortunes it encountered in common with many other property companies when the market collapsed were also examined.

Town and City probably built more buildings than any other property development company in the period 1956 to 1972. In the short space of some 15 years it was responsible for the development of literally thousands of shops and millions of square feet of offices and industrial space. In addition it made successful takeover bids for six major companies including Arndale, which was famous for its shopping malls. At the height of its career Town and City boasted assets of over £500 million. Like so many other companies whose directors were bent on expansion, it was necessary to secure a quotation on the London Stock Exchange if it was to gain prestige and attract both public and institutional finance.

Companies are always available which have a quotation but whose business for one reason or another has ceased to be profitable or sufficiently active to warrant their continuance. In the Far East there were companies whose tea and rubber plantations had either been sold off or perhaps confiscated. One such company was the Gan Kee Rubber Company Limited. This was acquired for a relatively nominal sum by Mr P.J. Goldberg, who previously ran Smart Brothers, the the multiple furnishing company, Mr J.B. Edwards a solicitor,

Mr Garvin an accountant and Mr J. Pollitzer of Beck and Pollitzer, the haulage contractors, a company heavy with cash following the nationalization of road transport.

The object of this company was to acquire and develop shop property, for these gentlemen had for some time been active in this field, doing their business with Messrs Jack Cotton and Partners, a well-known firm of estate agents originating in Birmingham but now active as well in the West End of London. Mr Barry East was a partner in this firm, through his connections with another partner, Mr Sam Messer (Mr Goldberg's cousin), with both of whom Mr East had been associated before the war and who were specialists in shop properties and investments arising out of them. Mr East, after six years service as a Captain in the Royal Engineers, was welcomed back to a different kind of experience, for Jack Cotton the senior partner was extremely active as a property buyer and developer on his own account. For the ten years following the war Mr East played a not insignificant part in the build-up of the Jack Cotton empire. It was not long before a conflict of interest became apparent since Jack Cotton, as a property developer, was not the only client of Jack Cotton and Partners, as a firm of estate agents. The new company which had now changed its name to Town and City Properties Limited relied upon their agents and, in particular, Mr East to expand their business.

The inevitable happened. Mr East was invited to devote his full time and talent to Town and City. With the prospect of the job of chief executive and with the opportunity to acquire substantial shareholding in the new company, he resigned his partnership with Jack Cotton and Partners and on 1 January 1957 at the age of forty-two he became chairman. Mr East, looking back on his earlier days, commented on the importance he put on his early training. He entered the profession of estate agent and surveyor at the age of seventeen, studied at night school and absorbed from his time in the army both the imposed and self discipline which he later translated into the administration of his business. He ventured the opinion that the younger generation of entrepreneurs lacking this training and, in particular, the disciplines were to some extent responsible for the wilder excesses that led to the recession in property. He saw and sees to this day, the image of a company as an extension of the personality and talents of the chief executive — in his own words 'a vehicle that moves at its best under a competent driver'. His pre-war experience and the ten years with Jack Cotton and Partners had provided Mr East with contacts with institutions such as the Prudential and the Legal and General which became the basis of the financial support of the company.

The first year of accounts shows a pre-tax profit of £15 000. Mr East recalls that the minutes of that meeting included the forecast that the profits might reach £200 000 per annum within ten years. In fact, in 1967, the profits rose to £1.3 million and, in 1973, no less than £5.4 million. In 1959, the Prudential Insurance Company called in Mr East to tell him that they were concerned at their inability to invest quickly enough the large revenues pouring into it. Traditionally property-oriented, the Prudential was looking for a developer through whom it could solve its particular liquidity problem. A compact ensued. A staggering £25 million was reserved for Town and City Properties Limited and this was to be only the first tranche. Town and City could call upon the Prudential to buy a site, pay for the new building to be erected on it and then lease it back to Town and City to show a return of 6%. Additionally, however, the Prudential was to get 20% of the growth of income at intervals of 33 years.

The package agreed between Town and City and the Prudential was that Town and City would not undertake a development to show less than 10%. Mr East quoted to me a kind of deal that worked time and time again. If a site cost £600 000 and the development costs were £400 000, totalling £1 million, then provided that the letting took place at £100 000 per annum, the ground rent would be £60 000 per annum and with no capital outlay at all, Town and City would benefit to the extent of £40 000 per annum.

This was not to be the end of the story, however. Initial rents were rising and Town and City could look upon their own reversionary prospects to show better results than was at first envisaged. They were able to look upon these top slices of income as becoming thicker and thicker as time went on, to the extent that the Prudential would accept them as security for loans to support other property purchases. Most of the original first tranche was used for shop development, but eventually offices and industrial properties were developed and Town and City moved into Europe and other overseas countries. The Prudential was content with these financial arrangements to such an extent that Town and City received what amounted to an accolade when Sir John Nellor Bt, in his annual statement in 1969 referred to the value that the Prudential placed on its association with Town and City. The policy of the company took a very definite shape; 40% of its development was in shopping, another 40% in offices and the remainder divided between industrial properties and those overseas.

The major decision as to what to buy, how the properties were to be designed, financed and managed, i.e., the very basis of the business, remained in the proven capable hands of the chairman and the original directors, Messrs Goldberg, Edwards and Garvin with Geoffrey Goodman as the advisory surveyor. Time, however, took its toll on all but the chairman, and due to retirements and deaths, Mr East was robbed of his original comfortable supporting team. Unfortunately, this was coincident with the increased problems that this by now huge undertaking had encountered by virtue of its size. The chairman found that it was necessary to set up a chain of command ensuring a great deal more delegation than had previously existed. A system of control was introduced based on local government procedures and later additions introduced methods favoured by American concerns. The personnel expanded and the chairman found himself more and more remote from the day-to-day decisions, many of which he had previously been involved in. More importantly, financial projections — previously the sole prerogative of the chairman, with assistance from Mr Garvin, the accountant (although he was generally brought in merely for checking purposes) — changed to elaborate charts, often presented in a manner which Mr East in retrospect feels was far less satisfactory than those for which he, alone, had been responsible. As in all large organizations, there is a tendency for certain individuals to set up small empires of their own; 'power delegated' Mr East reflects, 'is power lost'.

The chairman's strength was finance. He always wanted a long-term finance arrangement amounting to permanent finance of 100% and funded before the development took place. The volume of business before re-organization was such that relatively junior staff were engaged to investigate schemes and prepare viability *pro formae* originally drawn up by the chairman. These had to be approved by the finance directors and when they finally arrived at the chairman's desk, he was in a position to make sure 100% finance was available. Copies of the viability sheets were automatically sent to the Prudential Insurance Company in advance of the purchase contract for approval, ensuring no risk to Town and City.

In their anxiety to prove themselves, the new executives, creating as they did separate small empires of decision making, advocated new financing methods outside the Prudential pattern. They recommended merchant bankers only too keen to lend on short term to a company of the proven stability of Town and City. Mr East's explanation of the departure from proven financing methods is that he found that some of the deals involved assembling a site, which involved projecting a time-lag before the old-fashioned Prudential financial arrangement could be made. Mr East was persuaded that short-term borrowing was necessary to buy individual lots sometimes for very large amounts. Necessarily, the detail of site assembly must be a matter for conjecture at any time and the chairman was forced to believe his internal advisors, backed as they were by the readiness of bankers to provide finance at cheap rates. At one point in the history of the company, the total short-term indebtedness ran into £200 million.

Development was not only carried out by Town and City. The company acquired the following important property owning and development companies: Eldonwall, Arndale, London County and Midland, Sovereign Securities, Charlswood Alliance, Sterling Land, and Central and District Properties. Apart from these, Town and City purchased a number of investments in already constructed buildings, most of which were financed by the Prudential on the same basis as the development programme. With the exception of Central and District, the vast majority of these transactions were carried out before re-organization and were personally negotiated by the chairman.

In 1973 Town and City were offered the opportunity of acquiring the share capital of Central and District Properties

Limited for a sum of £97 million. Although this company owned commercial properties (including Berkeley Square House) its attraction to Town and City was the many potential sites contiguous with those already owned by Town and City and the huge reversionary prospects which were expected to mature within eight years. Mr East laid down his conditions for the acquisitions at this price:

1. Complete finance was to be made available.
2. Pre-sales to produce £70 million were to be arranged for 50% of the portfolio.
3. There was to be no significant impact on Town and City's pre-tax profits as a result of the shortfall on the remaining portfolio.
4. The financial arrangements were to include a roll-up of interest until the reversion expected in eight years' time.

Looking back on the matter, Mr East frankly regrets his decision to purchase. Whatever the results due to outside influences he considers that the decision was bad in principle, since it was completely out of character with all the policies he laid down so successfully in the past. Basically, this was a dealers' lot. He was influenced by its size, the estimate of £100 million profit on sales in reversion, and the spin-off of the sites contiguous to those owned by Town and City. Some idea of the possible profits can be gathered by considering Berkeley Square House which has some 400 000 square feet and at that time produced £1 million per annum or some £2.50 per square foot, but estimated to have a rental value of between £8 and £10 per square foot.

The event that followed was something just short of a catastrophe, once the decision to purchase had been made. The bankers agreed to lend the purchase price on fluctuating market terms, then at 8%. At a slightly higher rate the loan probably could have been fixed for a period of years but the bankers advised against a fixed rate as they firmly believed that interest rates were falling. Letters of agreement to purchase, but subject to contract, came from leading institutions for some £70 million of the property. Town and City committed themselves to the purchase. Mr East did not keep a diary of the dates when the blows fell, but rates of interest went up to 13%. Sales which had been agreed were aborted since the purchasers came under pressure and were unable to proceed. It was impossible to sell properties within the portfolio since the institutions, unsure of themselves and the property market, dared not buy even at the better yields offered. The properties in the portfolio, being reversionary in nature, were completely unsaleable in a market now subject to the rent freeze. With a threat of further legislation (and that from the Tory Government no less) the rent freeze iced up the lifeboat of the basic portfolio, preventing it from rescuing Central and District from the troubled waters in which it found itself and where it remained.

The Government edict that froze rents promised further legislation in connection with property without revealing what that legislation was to be. The effect on the property market was such that not only rents were frozen but the uncertainty as to the position of reversions froze values too. In many cases, the new higher yields that were now forced on the minds of institutions and their valuers threw up values considerably lower than the prices originally paid by many property companies. The boom that preceded these events was to a large extent one of finance and not construction. The ban on office development imposed by George Brown in 1964, and which affected not only London but Birmingham, Manchester and other centres mopped up any existing stock of empty buildings and furthermore had encouraged consumer demand probably in excess of real requirements. Commercial firms requiring say 10 000 square feet of space were told by their agents that buildings were running out in supply and that they should take, say, 15 000 square feet if the location was suitable as they would find themselves out in the cold if they did not settle on one of the limited buildings available. Many of these commercial firms found to their subsequent cost that they had fallen foul of that section of Parkinson's Law which states that empty offices will automatically find personnel to fill them.

The Yom Kippur War and the ensuing oil crisis had the effect of a general recession on trade. Expansion gave way to contraction. Expected reversions to higher rents now became a lesser consideration than getting any tenant to pay a rent at all. With a large number of schemes in the pipeline, many of them for sites which had not yet been fully pieced together,

Town and City Found itself with a potentially large development programme, the viability of which was now suspect. In March 1973, the pre-tax profit was £5.599 million, in 1974, this went down to £1.642 million, by 1975 there was a loss of over £12 million, and in 1976 a further loss of £4.75 million. The net assets of the company were down to £44 million. By 1976, it was, of course, necessary to sell as many completed investments as possible to retrieve the position.

The post-war period gave opportunities for individuals and very small groups who had the experience and talents to build up companies whose predominant image was one of outstanding financial success. For many of these companies and the individuals who directed them, the image was synonymous with the aspirations of the founders. This was clearly not the case with Mr East. Compared to the financial achievements of his company, his stake in it and the income he drew from it were modest. The quality of his buildings, confirmed by the status of the tenants he attracted to them was outstanding. The very size of the developments he undertook shows an imagination that clearly put him in the top rank of development specialists. Amongst the numerous and diverse developments that had taken place since the end of the war have been many that have aroused criticisms by town planning experts, architects and indeed the general public. No such criticism has been made of the developments carried out by Mr East, indeed his Arndale Centres and numerous provincial city centre developments are generally conceded to be the best of their type. Starting virtually from scratch he succeeded in building up the third largest property company in the UK. His talents were endorsed by the largest insurance company and, despite the severe setbacks to the company occasioned by a combination of outside influences, he deserves a special niche in the history of post-war development.

Before the conclusion of the interview, he laid down rules which he maintained should be observed by anyone running a company of this type:

1. Control must be kept of the very lowest level of administration at all times.
2. Borrowings or finance must be on a long-term basis.
3. Once a policy of finance and development procedure is

established and it is seen that this is successful, it is unwise to depart from such a policy.

4. Profitability and not size should be the motivation at all times.

Perhaps the fortunes of Town and City would have been different if Mr East had stuck to his declared policy of funding every development or purchase that was made in advance and on a very long-term or permanent basis. Borrowing short and lending long, whatever the possible profit involved, has always been a temptation likely to lead to financial embarrassment in banking circles. Property finance can be likened to this in many ways and there is not doubt that the motivation which was primarily to lead the field in size alone has played a great part in Barry East's career. The purchase of Central and District Properties was not so much an investment in development projects as a dealing or break-up situation far removed from the long-established pattern that had served Mr East so well.

Table 5.2 Town and City Limited movement of share price[10]

Date	*High*	*Low*
1965	25.05	19.66
1966	25.05	17.91
1967	31.50	19.85
1968	46.96	26.49
1969	48.44	32.56
1970	51.66	32.56
1971	79.70	41.05
1972	112.00	72.73
1973	115.00	50.00
1974	59.00	9.50
1975	35.00	8.50
1976	23.00	3.50
1977	15.00	5.00
1978	17.00	11.50
1979	25.00	14.00
1980	31.75	16.00
1981	29.25	27.25

Postscript

The company, after changing its name to Sterling Guarantee Trust PLC, merged, on 14/1/85, into P & O at an agreed share price of 74p — over 20 times its quoted price in 1976.

The main reason for the insolvent condition in which the company found itself must be attributed to rent control. By its nature the company relied for its profitability on the reversionary potential of site development and rent appreciation from properties already developed. The freezing of rents without a date for the thaw removed the certainty or near certainty of reversionary increases and, therefore, capital values, leading to a recall of the finance on which almost all property companies rely.

The story of Town and City Limited is by no means ended. The company still has vast assets although it has had to dispose of a number of its properties in order to reduce its indebtedness. It is no part of this study to prophesy the future of this or any other company, it is most unlikely that it will fail to retrieve its fortunes in time. It is likely, however, that without the guiding hand of a developer of the stature of Mr East it will not regain the dizzy heights it once reached in the hierarchy of the property company world (see Table 5.2).

5.6 TRANSPORT

5.6.1 Railways after World War II

During the war various schemes were put forward in which the Government was to assume responsibility for repairing war damage and to pay an annual charge to the railway companies. In February 1940 it was suggested that this charge should be £40 million per annum, later to be increased to £56 million per annum. In 1941 the scheme was finally worked out on the basis of the Government paying £43 million per annum and assuming the full risk of profit and loss. The cost of repairing was to be equally divided between the Government and the companies.

The Government showed a surplus over the years 1941—45 of some £200 million and this, no doubt, encouraged nationalization of the system under the 1947 Transport Act. The war years had taken their toll on unserviced locomotives and rolling stock, tracks and other equipment and it was felt that the restoration of the system would be beyond the resources of the companies. This, indeed, proved the case.

The cost of materials and labour rose more rapidly than receipts and the easing of petrol rationing exposed the competition of road transport. The Government thus found itself with an investment, the eventual capital cost of which was incapable of returning anything like a satisfactory yield. Each year showed an increasing loss and by the end of 1956 the deficit totalled £120 million, which the Government met by a special grant. Lord Beeching was made Chairman of British Railways; his task was to close many branch lines, replace steam by diesel and electric traction and to carry out improvements to a tune of £1500 million. In spite of all the efforts of the Government, British Railways still runs at a loss. It is probable that the continuance of any railway system must now be regarded as part of the defence of the country, of importance in times of war as a standby transport system for the armed forces or a national emergency. It is significant to observe that with the threat of a shortage of petroleum products affecting cars and lorries, the railway system which can be reconverted to consume coal instead of diesel fulfils a useful role.

5.6.2 Towns versus cars

This contest cannot be solved by mere bypassing since the centres of towns and cities are themselves focal points for the collection and delivery of both people and goods. Notwithstanding recent developments the majority of the towns and cities of Great Britain were designed with streets tailored to fit horse-drawn traffic, which was certainly slower, and for the most part in lesser number than the present-day motor traffic. To add to the congestion the majority of the buildings contain no garage space so that vehicles have to be left on streets, normally for longer periods than they spend moving along them. The average street is barely wide enough to accommodate three vehicles travelling abreast. Supposing that two of these vehicles are parked abreast but on opposite kerbs, this would mean that there is room for only one vehicle to pass between them. In spite of yellow lines and traffic wardens, the free passage of vehicles along many streets is hampered by all too common indiscriminate parking.

The effect on the centre of cities is the slow but sure

restriction of motor vehicles allowed to make full use of these centres. Some such restrictions are the introduction of parking meters and the escalating cost of using these, the rising costs of parking in car parks and the alteration in planning requirements. To further discourage unauthorized street car parking, the police have resorted to driving away vehicles and recently to clamping wheels with the 'Denver Boot', the removal of which takes time and costs out of all proportion to the convenience of parking in the chosen spot. The Greater London Council and its predecessor, the London County Council, at one time demanded a high rate of integral car parking in relation to the office content of any new building in London but in the last few years this policy has been reversed and whether the developer wishes it or not, car parking is restricted to a minimum in the hope of discouraging the use of cars in inner cities.

The effect of parking restrictions and the slowing down of traffic movement is observed in the growing pattern of changes in warehousing and distribution. Here the effect is two-fold. Multi-storeyed blocks in city centres used for warehousing are now going out of fashion. Obvious difficulties of loading and unloading mitigate against their future useful life and the changing pattern of retail distribution has severely curtailed the function of the erstwhile general warehouseman acting as a middle man between manufacturer and retailer. The growth of multiple firms, with many branches and dealing directly with manufacturers, necessitates their having points of distribution at convenient regional centres. These are better positioned outside busy town centres and are preferably sited on, or just off, motorways or trunk roads to give easy vehicular access.

Great Britain has not yet fully imitated the American and Canadian out-of-town shopping centres. Nevertheless, there are growing signs that despite opposition from local councillors, (who are very often local shopkeepers), high street concentration of shops, the conventional pattern for nearly a century, may well give way to newer shopping concepts where priority is given to car parking. However, we are yet to see the effect of hypermarkets with planned car parking set up in direct opposition to the traditional departmental stores.

Plate 9 Park West, Edgware Road, London. These buildings are typical of inter-war flat construction. (Courtesy of Press-Tige Pictures Ltd.)

Plate 10 'Spaghetti junction' — the road pattern in the environs of Birmingham. (Courtesy of Aerofilms Ltd.)

In a sense, the motor car can be regarded as part of the house, though it is free to come and go as its owner wishes. It is unlikely that it can be outlawed and governments of the future will have to live with it. If the number of motor vehicles grows, as past experience indicates will be the case, then two options are presented. Either the streets and roads through which the motor car has to travel must be made wider or, where this is not possible, the car must be banned from using them. It seems likely that a compromise will be reached, roads outside the main conurbations will almost certainly be provided in growing numbers to make inter-city travel as fast and unrestricted as possible. It seems likely that more and more streets in town and city centres will be closed to vehicular traffic, as is the case in such Central London shopping streets as Carnaby Street, South Molton Street and others. As far as town and city centres are concerned, the cost of acquisition of existing property to be demolished to widen roads would probably be too big a burden for the rate and income tax payers to bear. This is the theme of Professor Colin Buchanan's recommendations, so far as they affect existing urban centres where the capital costs involved make drastic traffic improvement by street widening impossible. Where new towns are planned, Buchanan advocates much closer attention to planning the location of buildings for various users to anticipate prospective traffic flow.

5.6.3 Vertical transportation

An important consideration that must not be overlooked is the increased density of city buildings. To a large extent, this has been made possible by the development of the lift. Although in Roman times, as early as 236 BC, Vitruvius described devices for lifting weights it was not until the seventeenth century that a passenger lift was developed, operated by hand power through a system of counter weights. The advent of electric lifts came in the late nineteenth century, being developed in Germany and America, in the latter by Otis in 1889. Modern technology has increased the efficiency of lifts and enabled heights exceeding 300 feet to be easily operable. The imposition of plot ratios by the planners prevents the commercial viability of a development from benefiting fully as a result of building upwards, beyond an optimum height. The higher the building rises the smaller becomes the individual floor area and the ratio of gross built area to nett lettable area increases.

Most towns in the United Kingdom were developed with streets far narrower than those found in America so that taller buildings in the latter case do not obstruct the natural light of those on the other side of the street. This and other such aesthetic reasons are responsible for restricting building heights and, therefore, the maximum use to which lifts can be put.

REFERENCES

1. Crossman, R. (1975) *The Diaries of a Cabinet Minister.* Hamish Hamilton and Jonathan Cape, London.
2. Cherry, G. E. (1974) *The Evolution of British Town Planning.* Leonard Hill, Leighton Buzzard.
3. Cullingworth, J. B. (1975) *Environmental Planning 1939—1959.* HMSO, London.
4. Hall, P. (1973) *The Containment of Urban England.* Allen and Unwin, London.
5. Hepker, M. Z. and Whitehouse, C. J. (1975) *Capital Transfer Tax.* Heinemann, London.
6. Friedman, M. and Stigler, G. J. (1972) *Roofs and Ceilings: the Current Housing Problem.* Institute of Economic Affairs, London.
7. Pennance, F. G. (1972) in *Verdict on Rent Control* (eds F. A. Hayek *et al.*). Institute of Economic Affairs, London.
8. Paish, S. W. (1972) Recent British experience: a postscript form 1975, in *Verdict on Rent Control* (eds F. A. Hayek *et al.*). Institute of Economic Affairs, London.

FURTHER READING

Ambrose, R. and Colemitt, B. (1975) *The Property Machine.* Penguin, Harmondsworth.

Bor, W. (1972) *The Making of Cities.* Leonard Hill, Aylesbury.

Buchanan Report, The (1964) *Traffic in Towns.* Penguin, Harmondsworth.

Burke, G. (1976) *Townscapes.* Pelican, Harmondsworth.

Chrisfield, L. (1983) *Taxing the Profit.* Centre for Advanced Land Studies, Reading.

Cullingworth, J. B. (1976) *Town and Country Planning In Britain.* Allen and Unwin, London.

Darlow, C. (ed.) (1982) Valuation and development appraisal. *Estates Gazette*, London.

Darlow, C. (ed.) (1983) Valuation and investment appraisal. *Estates Gazette*, London.

Davis, R. L. and Champion, A. G. (eds) (1983) *The Future for the City Centre.* Academic Press, London.

Department of Environment (1976) *Transport Policy — A Consultation Document.* HMSO, London.

Department of Environment Property Advisory Group (1983) *The Climate for Public and Private Partnerships in Property Development.* HMSO, London.

Foster, C. D. (1975) *The Transport Problem.* Croom-Helm, Beckenham.

Heap, D. (1967) *Introducing the Land Commission Act.* Sweet and Maxwell, London.

Heap, D. (1973) *An Outline of Planning Law.* 6th edn. Sweet and Maxwell, London.

Hepker, M. Z. (1973) *A Modern Approach to Tax Law.* Heinemann, London.

Holliday, J. (ed.) (1973) *City Centre Redevelopment.* Charles Knight and Co., London.

Maas, R. W. (1976) *Development Land Tax.* Tolley, Croydon.

Marriott, O. (1976) *The Property Boom.* Hamilton, London.

Matthews, J. and Johnson, T. A. (1977) Development land tax. *Estates Gazette*, London.

Mellows, A. R. (1982) *Taxation of Land Transaction.* Butterworths, London.

Nash, C. A. (1976) *Public versus Private Transport.* Macmillan, London.

Partington, M. *Landlord and Tenant.* Weidenfield and Nicolson, London.

Pilcher, D. (chairman) (1975) First report of the Advisory Group on Commercial Property Development to the Secretary of State for the Environment. HMSO, London.

Ratcliffe, J. (1974) *Introduction to Town and Country Planning.* Hutchinson, London.

Sadikali, M. (ed.) (1983) *Butterworths' Orange Tax Handbook.* 8th edn. Butterworths, London.

Sadikali, M. (ed.) (1983) *Butterworths' Yellow Tax Handbook 1983–4.* 22nd edn, with supplement. Butterworths, London.

Soars, P. C. (1982) *Land and Tax Planning.* Oyez Longman, London.

Short, J. R. (1982) *Housing in Britain — The Post War Experience.* Methuen, London.

The Structure and Activity of the Development Industry. A memorandum to the Property Advisory Group (1979) Royal Town Planning Institute, London.

Telling, A. E. (1977) *Planning Law and Procedure.* Butterworths, London.

Tiley, J. (1983) *Butterworths' U.K. Tax Guide 1983–4.* 2nd edn. Butterworths, London.

6

Conclusions

6.1 INTRODUCTION

G. M. Trevelyan[1] concluded his *Shortened History of England* with the words

> 'In the earlier age man's impotence to contend with nature made his life brutish and brief. Today his very command over nature so admirably and marvellously won has become his greatest peril. Of the future the historian can see no more than others. He can only point like a showman to the things of the past with their manifold and mysterious message.'

The philosophical thought which engenders the reality is clearly evident both in the teachings of the earliest sages and those that have followed to the present time. The issue can be stated simply as the struggle to determine who controls and profits from the ownership of property. Philosophy, economics and politics are the three sides of a triangle, the lengths of which will vary, seldom appearing equilateral. The shape of the triangle was governed from earliest times; the longest side in the first half of the nineteenth century was economics. In the latter half of the nineteenth century Governments were forced to enact remedial legislation. This was in retrospect the least that any civilized society could do to improve conditions before the health and, therefore, the efficiency of much of the working population were ruined. Most of the legislation applied to places and conditions of work but little was done to provide better housing and such as emerged was achieved by private enterprise virtually uncontrolled. Local authorities were given a mandate to clear the worst of the slums but not to replace them with new houses. It was not until the last decade of the century that improvements in transport were followed by the growth of inner suburbs offering better housing.

The antagonism between employers and employees, many of the former property owners and the latter tenants, was inflamed by the conditions arising directly from the excesses of the Industrial Revolution. The concentration of workers in certain cities, where previously they had been scattered in a largely agricultural environment, undoubtedly revived ancient philosophical attitudes leading to Socialism and in particular, the part that property should play in the social contract that was always implied if not specifically chronicled. The first years of the twentieth century witnessed the attempt by Lloyd George to control the profits arising out of ownership but not the use to which it could be put. The first Planning Act of 1909 was to encourage a more orderly pattern of development but was not mandatory. The 1915 rent control legislation was a measure of expediency, the preamble to the Act stating it was to operate for the duration of the war and to continue for six months only thereafter. The results were the repeal of the first and the ineffectiveness of the second. The third sowed the seeds of a plant whose foliage has overgrown and virtually obliterated the rented residential market but which, more importantly, has been indirectly responsible for the diversion of the vast accumulation of funds representing the savings of the bulk of the population into commercial property and away from investment in industry and government securities.

Before 1914 planning was voluntary, as is evidenced by the paucity of legislation that emerged even after the recommendations of committees set up during the First World War. Those of the Second World War, however, provided the base for the 1947 legislation which attempted three goals, the redistribution of industry evenly throughout the country, an orderly pattern of local development by control of use and the annexation by the State of profits arising from it.

Events have shown that the policy of the distribution of industry has been frustrated by changes in the pattern of industry itself. The heavy industries that once were the backbone of the economy have given place to others, such as electronics, and the numbers employed in offices have greatly increased. Notwithstanding attempts to legislate for a halt in the expansion of the South East in 1964 the repeal of the Brown ban has been followed by the continuance of the trend by industry to locate independently, despite Government exhortation or decree.

Population increases and the demand for better housing standards was met inadequately in the latter half of the nineteenth century by a few industrialists producing village garden estates in the provinces and 'improved dwellings' (principally tenement blocks in London and Glasgow for example). Government sponsorship of housing between the wars enabled private enterprise to achieve a surplus. Rent control after 1945 was a disincentive to the development of rented housing and moreover directed its provision by municipalities. Labour Governments have thought that council tenants are likely to give them their votes and the reaction by Conservative Governments is to sell to tenants hoping that home ownership creates a more capitalistic and, therefore, Conservative society.

Much of the legislation passed by both political parties has been confused and divergent and in many instances has not produced the intended results. Politicians do not fully understand the business of property. Even without the restraints and impositions they have created to control its development, use and profitability would have been achieved by allowing purely market forces to act. With a few exceptions the property business was principally in the hands of individuals or small groups until the 1950s. From thereon the numbers of publicly quoted companies increased and the institutions changed their role from money-lenders to proprietors and developers. The change and the vast increase in the investment of their funds has altered the manner in which the business is conducted including research, computerization and a managerial system as opposed to the 'one man band' operation that characterized the post-war booms.

Planning in the true sense of the word should be concerned with the orderly development and use of property, both at the local and national stage. Apart from the few new towns where planning was positive, the redevelopment of existing towns has shown up the planners' exercising powers of veto through the system which allows them excessive time delays in approving plans, acting as arbiters of taste, all best described as negative planning. The interaction of taxation and planning has bedevilled the property industry since the Lloyd George proposals reintroduced in 1947, and has resulted in the enactment and repeal of at least three attempts to levy special taxation on property development. The development land tax that still remains on the statute books is the one example where a commercial venture incurs a tax on its presumed profit before this is made.

6.2 RENT CONTROL AND SECURITY OF TENURE

In the commercial sector much of the antagonism that might have occurred between landlords and tenants has been avoided by the provisions of Part II of The Landlord and Tenant Act of 1954 which gave security of tenure to tenants and provided for a fair market rent to be paid to landlords. In the residential sector the piecemeal legislation since 1945 has resulted in the virtual disappearance of the private rented sector. This has permitted the dereliction of areas of low rented accommodation where inadequate rents discourage landlords from effecting repairs and dissuade investment in new housing for renting. The difference in price between a house with vacant possession and one let on a controlled rent means that the landlord subsidizes the tenant. If control was removed rents would rise. A tenant unable to pay would either seek a higher wage or State assistance. It is inconceivable that a Labour Government would repeal control and it is possible that a Conservative Government would prefer the continuance in the decline of rented property and the emergence of a society of owner-occupiers. The role of the building societies and banks who now are prepared to lend almost 100% of the purchase price of a house tends to suggest that the future of the residential sector is likely to follow this trend.

6.3 TAXATION OF CAPITAL

Changes in taxation not directly aimed at property have nevertheless affected non-institutional property companies and individuals. The most important of these is the replacement of death duties by the Capital Transfer Tax and the Capital Gains Tax. Together they have prevented the wealthier property owners or shareholders in property companies from creating trusts in their lifetime or bequeathing assets to their heirs. Thus after death tax will compel sales, most likely to institutions. These taxes, of course, apply universally; in the property industry, however, their effect is likely to be a periodic disappearance of the smaller older companies, (as they are swallowed up by institutions), and their replacement by new companies set up by emerging young entrepreneurs, which in turn, will suffer the same fate.

6.4 LOCAL TAXATION

Local taxation in the form of rating, dating back to the reign of Elizabeth I was designed to do no more than provide for the poor of the parish so as to prevent their wandering from place to place. In modern times, rates are levied to defray the expense of many items of a regional if not national character. They fall heavily on 'commercial' individuals and companies who have no voting power: only residential occupiers are entitled to a say in borough and county elections. Recently proposed legislation to regulate the spending of local authorities will help to prevent higher than present charges but the system is too entrenched to allow for a change to local purchase taxes or an increase in income or corporation tax. The occurrence of adjoining boroughs with one charging higher rates than its neighbour creates the danger of the business rate payer moving the short distance to get the better bargain, or staying but demanding a cheaper rent or price to compensate for the higher rate bill.

In its favour it is a tax whose amount is easy to calculate and collect. Whatever its faults, the cost and disturbance involved in changing it have persuaded successive Governments to retain the system.

6.5 POWERED TRANSPORT

Changes in modes of transport have, in the main, affected more the location of centres of urban development than the centres themselves. Such legislation that has occurred has been of economic incidence with little or no opposition from opponents of the Government in power at the time. There is perhaps one exception, that of the repeal of the Road Transport Act enacted by the first post-war Labour Government. Certainly, the opening of a new railway station has had a dramatic effect on the expansion of the area surrounding it and one such case is dealt with in some detail in the text. Of some significance is the interaction of transport and speculative building, exemplified in the suburban expansion of London between the turn of the century and the 1930s, but this may be considered as an isolated occurrence unlikely to be repeated.

The development of new forms of transport throughout the nineteenth and twentieth centuries has decreased the popularity of the preceding modes of travel: witness the virtual disappearance of the horse, the tramway and the canal as viable means; the decline of railway profits (necessitating nationalization in order to retain the system and the municipalization of inner city buses.

It is the motor car that has had the most profound effect on urban development, chiefly by the sheer weight of numbers, demanding not only the room to manoeuvre but to remain stationary when not in use. Statistical evidence would seem superfluous to support the observation of any eye on the various effects of the car on the urban scene. Attempts to curb the free use of roads and streets are so far confined to a few pedestrianized shopping areas. In the field of office and shop development, planning policy initially insisted on adequate parking spaces to be provided within the curtilage of a new development. Overall planning policy changed to a virtual ban on car-parking space in order to discourage the use of cars between the home and the place of employment. To date this latter policy predominates. This example is of minor significance compared to the major problem of adapting urbanism to the ever-increasing number of cars. This requires, according to Professor Buchanan, major changes of such proportions both in road patterns and in redevelopment of existing street patterns that they are likely to be at a cost beyond the present resources available to the Government.

It is possible, however, to observe the immediate effects of the pressures brought to bear by the weight of motor car traffic. With restrictions on parking and the difficulty of loading and unloading goods, warehouses which were formerly located in town centres, usually near railway termini, have been found uneconomic and thus there has been a tendency for relocation on or near major systems outside city or town centres. To some extent this exodus has been copied by offices, although other considerations discussed elsewhere have contributed to decentralization. Almost certainly the relocation of industry from urban centres has followed the trend and may be traced to the roll-on effect of suburban development following improved transport facilities provided by rail, tramways, bus and underground. These facilities were intended to bring the work force to the centralized factories and offices. Now with congested centres, the attractions of moving works storage and offices to the suburbs and beyond are being pursued. Thus, the improvement in transport has had a seemingly reverse effect resulting in, for example, the growth of Croydon as an office centre (although the fast train service of some fifteen minutes duration to Victoria offers some explanation). Equally significant is the not unexpected growth of office development at Slough, Windsor and Reading as a direct result of their proximity to Heathrow Airport.

The interaction of suburban growth and the transport facilities that either led or followed this, set a standard of housing accommodation higher than that of the nineteenth century; rent restriction on building maintenance has led to the deterioration of large numbers of housing units in inner urban areas. Planning requirements have militated against the demolition which, together with reduced demand of equally outmoded manufactured products and services, has added to the decline of certain pockets of inner city areas. An outstanding example is the dockland area of London and a similar effect is apparent in Liverpool and Glasgow, the result of a change in the transport of sea-borne goods by the container system and a world-wide decline in the shipping industry. The dereliction of both residential and commercial property in these areas has been followed by plans for redevelopment, but no lobby has appeared calling for these areas to be planned as open spaces as compensation for the suburban and rural areas of open spaces developed in advance to replace them.

One form of transport has not been discussed — the individual's ability to walk. Whilst the location of property has been affected by differing forms of aided transportation it may be confidently concluded that so long as men resolve to meet together for any purpose, social or commercial, rather than communicate by telephone or electronic means, there will be market centres such as the Stock Exchange, Lloyds and the Baltic Exchange in London, with similar if fewer examples in the major towns and cities. It would seem unlikely that mechanical or electronic devices will replace the need for markets such as these, and the myriad of other commercial activities where physical contact is considered essential, is unlikely to result in the decline of city centre location of business.

If this proposition is valid the world of property is left with the problem of communication by transport, a continuing subject for discussion and action. It is hoped that a compromise solution will be reached, rather than a ban on all but carefully controlled public transport in towns, with a consequent reduction in the number and use of private cars.

Plate 11 The Brent Cross shopping mall, London—the style of the enclosed shopping centre of the future. The illustration does not show the large car park forming part of the development (Courtesy of Press-Tige Pictures Ltd.)

Plate 12 St Katharine's Dock, East London, part of the vast redevelopment scheme replacing the outmoded docklands. (Courtesy of Press-Tige Pictures Ltd.)

6.6 THE ROLE OF PROPERTY IN THE ECONOMY – A SUBJECT FOR FURTHER RESEARCH

In separating the building boom of 1954—64 from the financial boom of 1970—73, it may be concluded that an underlying malaise of investment in property rather than industry has been present throughout the entire post-war period, the germ of which has lain dormant in the body of the entire economy. As a result of the financial boom of 1970—73, a widely held conviction was created, partly by the media but to a not inconsiderable extent by some of the participants themselves. This was that involvement in the property business was a passport to inordinate wealth created with the minimum of effort or skill by those least entitled to the rewards and at the expense of the 'community'. Even otherwise intelligent and well-informed commentators voiced such views and inevitably, either as a result of this pressure, or more probably recognizing a ready-made scapegoat, the Government of the day enacted repressive and punitive tax legislation. They introduced lending limits to the property sector and not necessarily specifically for that purpose and increased lending rates, all of which contributed to the 'crash of 1974'. In the inquest that followed, the Heath/Walker/Barber expansionist philosophy was held to be the culprit, aided and abetted by rapacious secondary and tertiary banks and other lending agencies, all of which it was said, provided the catalyst to the potentially dangerous compound of avaricious capitalists, exploiting the opportunities provided by a consumer-led economy.

That was the scenario which was widely accepted at the time, but subsequent events disclosed that perhaps, after all, things were not exactly as they appeared. Two individuals who reputedly made £18 million profit from the sale of Bush House, Aldwych, to a nationalized industry pension fund, were not after all the multi-millionaires they were thought to be. Gabriel Harrison, chairman of the publicly quoted Amalgamated Properties, died leaving a net estate of virtually nil. William Stern entered the *Guiness Book of Records* with a personal bankruptcy in excess of £100 million, while the subsequent enquiry into the Peachey Corporation showed the chairman, Sir Eric Miller, only sustained his lifestyle and friendship with those in the highest places by questionable actions. Even the inventor of the OPM (other people's money) theory saw the total disappearance of the Slater Walker Empire and must at times have wondered whether his next overseas trip would have been as a guest of the Government of Singapore; the list of the mighty that fell is a matter for the record. In the circumstances the cases may differ in some respects, however the totality of the events of 1974—76 was the virtual destruction of the secondary banking system as it was known in the post-war era, the reduction from over 110 to just over 80 publicly quoted property companies, the personal bankruptcy and in some cases disgrace of property men previously held in the highest esteem, and the loss to shareholders of hundreds of millions of pounds. Could it really be true that these immense sums of money were, in fact, earned in the years of the financial boom and lost overnight, or was it after all only a figment of the imagination?

At present a detached judgement of events is probably still not possible but in years to come it may well be concluded that in reality the property fortunes were made in the building boom of 1955—70 and not in the so-called financial boom. It may be judged that in 1970—73, rather than fortunes being made, what was really created were the conditions necessary to precipitate a catastrophic collapse of one of the props upon which the economic structure of the country rested.

The expansionist policy of the Heath administration was well understood and even the protagonists of the theory conceded that it represented something of an economic gamble. It undoubtedly led to a very rapid expansion of the money supply with the consequence that opportunities occurred for borrowing money at relatively acceptable interest rates. A total lack of vigilance on the part of the authorities also enabled the setting-up and expansion of lending agencies unfettered by the restraints inherent in the primary banking system. The City of London's banking system, developed over a period of more than a hundred years, was pre-eminent throughout the world both for its integrity and sagacity. Banks failed in South America or East of Suez, not in London.

However, the availability of money in itself clearly does

not create a demand for one particular commodity to the total exclusion of all others unless some other factor intervenes to produce such inexplicable selectivity. In fact, two other elements were present to create the conditions which triggered off the financial boom. Both of these conditions have passed largely unnoticed, only the first of these has engendered any comment, and that only recently by stockbrokers specializing in property shares. Contrary to universally accepted practice, it became increasingly fashionable from 1960 onwards to value quoted property company shares not on a comparative price : earnings ratio but on net asset value. Investors apparently were disinterested in income and looked only for potential gain. Given this growing inclination of the investing public, the quoted property company provided the total answer. Unlike companies in almost all other sectors, the value of the assets required certification by specialist valuers not, as in other cases, by the established auditors to the company. Banks and finance houses measure their assets by reference to cash deposits, secured debts and discounted bills, whilst manufacturing and trading companies have tangible stocks and work-in-progress, the value of which can be definitively measured. Conversely, the income of a property company comprises rents which are finite and exactly comparable to the earnings produced by any other company by way of sales, fees or commissions.

It is not, therefore, difficult to envisage the opportunities afforded for the escalation of the value of property companies once the conventional price : earnings basis of valuation is abandoned. The gross rental income is factual, the capital valuation of the property is in the mind of the valuer and if not the first valuer, then there is always another! Was then the expansion of the money supply and its ready availability through the burgeoning secondary banking system, coupled with a reassessment of the value of property companies, collectively sufficient to trigger off the boom? The answer must be 'no'. The enormity of the consequences only narrowly averted by uncharacteristically prompt and effective Government action in establishing the Bank of England lifeboat with resources of £1000 million dictates that a search must be made for a fundamental underlying economic *raison d'être* for the events.

During the slow and painful recovery of the property sector in 1976–79, interest rates were as high if not higher than the disaster years 1974–75 and the same economic malaise, static industrial production, high levels of unemployment and chaotic industrial relations subsisted through both periods; yet one was the aftermath of a disaster whilst the other signalled the gradual but certain recovery of the sector. The boom, the crash and the slow recovery — are these all perhaps an illusion? Are we in fact, without knowing it, looking at a continuing remorseless process in which the apparent disaster was no more than a minor accident which, whilst temporarily shattering confidence, will in fact fail to halt or divert the course of events? If that is the case, what is the force which remorselessly fuels this section of the economy?

The Wilson committee on the workings of the City heard evidence of disquiet among the Stock Exchange fraternity regarding the proportion of equity in UK quoted companies owned by the institutions. An indication of the net spending on property by institutions since 1963 is given in an article entitled 'Commercial property investment in the next five years', by Michael Harris,[2] which indicated 'between 1963 and 1978 spending on property increased from £175 to £1,330 million, an increase of about 7½ times or a compound annual increase or around 13%'.

The stockbrokers' anxiety, however, is based on their fear that the institutions' power is increasing to the point that they can collectively, if not individually, control the market and thus may seek to establish a system of equity dealing which would by-pass the established exchanges. While this sectional interest is understandable, it is surprising that greater emphasis is not placed upon the virtually non-existent new capital commitment by the institutions to productive industry, whilst at the same time the annual total invested by the institutions in real property grows at an ever-increasing pace. The investment by institutions in real property will, by 1985, make the total investment by the much maligned and now deceased banks seem like petty cash. Were the secondary banks after all not the villains of the piece but the scapegoats? Was the Heath/Barber economic philosophy the cause of the boom and, therefore, the crash? Spectacular as was William Stern's debt of £100 million to the Crown

Agents in the context of a personal guarantee, was it a significant sum relative to the sector as a whole?

It is estimated that 75% of the adult working population in the United Kingdom is entitled to retirement pensions from occupational pension funds. The vast majority of these pensions are related in some way to earnings in the last few years of employment. Thus it becomes apparent that with continuing inflation, the liability on the funds to provide pensions increases by geometric rather than arithmetic progression and the constant quest of the fund managers and actuaries is for a haven for current contributions where the monetary performance will match the ever-increasing ultimate demands upon the funds. Incredibly at first sight, yet understandably, none of the funds show any great inclination to invest capital in the industries which produce the earnings from which the pension contributions are derived. Have the miners invested in British Leyland, the airline pilots in steel strip mills in South Wales, or the electricity workers in the shipyards of the Clyde? No, the fund managers have their job to do and a goodly proportion, increasing each year, will go where it has gone for the last ten years — into real estate. Is this then because over the post-war period property has performed better overall than any other available investment in the United Kingdom? This is a complex and perhaps impossible question to answer. The instinctive answer from those both inside and outside the property world will be in the form of clichés: 'there is nothing better than land, they have stopped making it', 'you can't beat bricks and mortar', 'safe as houses', etc., but all comparisons are relative and the options open to the pension funds must be viewed in the light of their virtual tax exempt status. That is to say tax exempt on their total receipts, whether they be in the form of dividends from shareholdings in dividend paying companies or by way of interest on mortgages or rents from properties held. That is not to say that the companies paying dividends to pension funds are themselves exempt from normal corporation tax upon the profits earned contributing to those dividends. Neither does this immediately explain the effect of pension fund investment on the attraction of property shares to others. This latter situation arises by virtue of the fact that property companies engaged in development have, on completion and letting of their developments, a ready-made purchase in the form of the pension fund competing for prime property with effectively tax exempt money. The knock-on effect of pension fund participation in the market is to improve the profit of the property company. It enables it to service increased borrowings, thus giving purchasing power for further properties to be valued with a weather eye for asset value rather than income and the ingredients of a boom are already in the mixing plan.

But the cliché stating that 'they have stopped making land' contains an element of truth. There is a finite limit to the amount of built real property which can be supported by the economy in any country. The degeneration and decay of many inner city urban areas is not entirely the result of misguided planning principles, it has only been exacerbated by unintelligent policies dictated by political dogma. Essentially it is symptomatic of an economic process which commenced with the Industrial Revolution, has continued remorselessly since, obscured by two World Wars, but is about to accelerate beyond any previous experience with the advent of the microchip industrial revolution of the 1980s. The truth is that there is a gross surplus of urban built property. Leaving domestic housing out of consideration, it is arguable that as much as 30% of built property is now obsolescent and the percentage looks like increasing during the next decade. The frightening possibility is that there will be no necessity, no demand, no reason to replace other than a very small proportion of the growing obsolescent total. Yet the taxation system relative to pension funds, the political-social attitude to financial support during pensionable age, the conviction that thirty years of attending a place of work entitles the individual to be supported by the ever-decreasing number of productive workers in the country for the rest of his natural life, (in itself prolonged by progressively successful medicine) must inevitably bring about a total breakdown of the current concepts of institutional investment in real estate.

If and/or when that occurs it could make the crash of 1973 pall into insignificance. Is this the time to call a halt? Advanced social economic theories in the current political climate require many years of discussion, evolution and testing. Judged by their performance, the economic grasp of our elected legislators is minimal if not non-existent, although it must be recognized that in dealing with fundamentals, regard must be given to what is politically achievable. Fiscally, however, it is within the power of a majority Government at the commencement of a five-year term to enact unpopular, even virtually unacceptable, measures provided that during its term it can placate the electorate by other means.

The one fiscal measure that a bold Government could enact would be to make all 'gross fund institutions' pay corporation tax at precisely the same level as all other tax-paying bodies. It is not within the scope of this book to advance the ways and means by which corporation tax should apply to all at substantially lower levels than are prevailing at the present time.

By tax concessions to companies, particularly nationalized industries, successive Governments have encouraged the means to provide pension funds to prop up staff schemes which, in themselves, could never provide pensions at acceptable levels. The inevitable diversion of massive funds from investment in productive industry to so-called 'safe property' has contributed to the decline in competitiveness of British industry, although this has been offset by a temporary salvation due to North Sea oil. There should be a fundamental realignment of national investment before it is too late. It is wholly unrealistic to believe that 20 million workers can be provided with pensions at the promised levels from property rental income derived from the collection of tenants comprising British industry and commerce, many of whom are in financial straits if not already bankrupt.

REFERENCES

1. Trevelyan, G. M. (1970) *Shortened History of England*, Penguin, Harmondsworth.
2. Harris, M. (1980) *The Chartered Surveyor*.

Index

For Product Safety Concerns and Information please contact our EU representative GPSR@taylorandfrancis.com Taylor & Francis Verlag GmbH, Kaufingerstraße 24, 80331 München, Germany

Printed and bound by CPI Group (UK) Ltd, Croydon, CR0 4YY
11/04/2025
01843977-0007